The Potential Distribution Theorem and Models
of Molecular Solutions

An understanding of molecular statistical thermodynamic theory is fundamental to the appreciation of molecular solutions. This complex subject has been simplified by the authors with down-to-earth presentations of molecular theory. Using the potential distribution theorem (PDT) as the basis, the text provides an up-to-date discussion of practical theories in conjunction with simulation results. The authors discuss the field in a concise and simple manner, illustrating the text with useful models of solution thermodynamics and numerous exercises. Modern quasi-chemical theories that permit statistical thermodynamic properties to be studied on the basis of electronic structure calculations are given extended development, as is the testing of those theoretical results with *ab initio* molecular dynamics simulations. The book is intended for students undertaking research problems of molecular science in chemistry, chemical engineering, biochemistry, pharmaceutical chemistry, nanotechnology, and biotechnology.

THOMAS L. BECK is Professor of Chemistry and Physics at the University of Cincinnati. His research focuses on the development of quantum simulation methods and solution phase theory and modeling.

MICHAEL E. PAULAITIS is Professor of Chemical and Biomolecular Engineering and Ohio Eminent Scholar at Ohio State University. He is also Director of the Institute of Multiscale Modeling of Biological Interactions at Johns Hopkins University. His research focuses on molecular thermodynamics of hydration, protein solution thermodynamics, and molecular simulations of biological macromolecules.

LAWRENCE R. PRATT works in the Theoretical Chemistry and Molecular Physics Group at Los Alamos National Laboratory. His research focuses on theoretical problems in chemical physics, especially on the theory of molecular solutions and on hydration problems in molecular biophysics.

THE POTENTIAL DISTRIBUTION THEOREM AND MODELS OF MOLECULAR SOLUTIONS

THOMAS L. BECK

Departments of Chemistry and Physics, University of Cincinnati

MICHAEL E. PAULAITIS

*Department of Chemical and Biomolecular Engineering,
Ohio State University*

LAWRENCE R. PRATT

Theoretical Division, Los Alamos National Laboratory

CAMBRIDGE
UNIVERSITY PRESS

CAMBRIDGE UNIVERSITY PRESS
Cambridge, New York, Melbourne, Madrid, Cape Town, Singapore, São Paulo

Cambridge University Press
The Edinburgh Building, Cambridge, CB2 2RU, UK

Published in the United States of America by Cambridge University Press, New York

www.cambridge.org
Information on this title: www.cambridge.org/9780521822152

First published 2006

Printed in the United Kingdom at the University Press, Cambridge

A catalog record for this publication is available from the British Library

ISBN-13 978-0-521-82215-2 hardback
ISBN-10 0-521-82215-7 hardback

Contents

Preface

Molecular liquids are complicated because the defining characteristics that enliven the interesting cases are precisely molecular-scale details. We argue here that practical molecular theory can be simpler than this first observation suggests. Our argument is based upon the view that an effective tool for developing theoretical models is the potential distribution theorem, a local partition function to be used with generally available ideas for evaluating partition functions. An approach based upon the potential distribution theorem also allows functional theory to ride atop simulation calculations, clearly a prudent attitude in the present age of simulation.

This work is about molecular theory, and emphatically not about how to perform simulations. Molecular simulation is an essential component of modern research on solutions. There are a number of presentations of simulation techniques, but not of the molecular theory that we take up here. We offer this book as complementary theory with simulators in mind.

A goal of this book is, thus, to encourage those performing detailed calculations for molecular solutions to learn some of the theory and some of the sources. The physical insights permitted by those calculations are more likely to become apparent with an understanding of the theory that goes beyond the difficulties of executing molecular simulations. Confronting the enormity and lack of specificity of statistical mechanics usually would not be the practical strategy to achieve that goal.

This book also frequently attempts to persuade the reader that these problems can be simple. Extended discussions are directly *physical*, i.e., non-technical. This is consistent with our view that many of these problems *are* simple when viewed from the right perspective. Part of our view is, however, associated with a high-altitude style: in many instances, we are comfortable in presenting things simply and referring to comprehensive sources for background details (Münster, 1969,1974).

Thus, an introductory course in statistical thermodynamics, typically offered in graduate programs in chemistry and chemical engineering, is a prerequisite to this material. A few sections are at a level more advanced than that. But the references and access to a technical library, or to a knowledgeable teacher, would provide the natural supplement. We hope that this book will be accessible then to students with a strong background in a physical science, and specifically to graduate students embarking on research activities in molecular modeling of solutions in chemistry, chemical engineering, biophysics, pharmaceutical sciences, and in molecular biotechnology and nanotechnology.

We have made a conscious decision to emphasize aspects of the theory of molecular liquids different from the mature and familiar theory of atomic liquids. This decision is partly due to the fact that the theory of simple liquids is well described elsewhere, and partly due to the view that the specifically *molecular* aspects of solutions are essential to topics of current interest.

It is helpful to contrast the view we adopt in this book with the perspective of Hill (1986). In that case, the normative example is some separable system such as the polyatomic ideal gas. Evaluation of a partition function for a small system is then the essential task of application of the model theory. Series expansions, such as a virial expansion, are exploited to evaluate corrections when necessary. Examples of that type fill out the concepts. In the present book, we establish and then exploit the potential distribution theorem. Evaluation of the same partition functions will still be required. But we won't stop with an assumption of separability. On the basis of the potential distribution theorem, we then formulate additional simplified low-dimensional partition function models to describe many-body effects. Quasi-chemical treatments are prototypes for those subsequent approximate models. Though the design of the subsequent calculation is often heuristic, the more basic development here focuses on theories for discovery of those model partition functions. These deeper theoretical tools are known in more esoteric settings, but haven't been used to fill out the picture we present here.

Exercises are included all along, but not in the style of a textbook for a conventional academic discipline. Instead we intend the exercises to permit a more natural dialogue, e.g. by reserving technical issues for secondary consideration, or by framing consideration of an example that might be off the course of the main discussion.

The Platonic debate known as "The Learner's Paradox" suggests *if you don't know it, you can't learn it.* A judgment on the truth content of this assertion is tangential to the observation that learning is difficult in ways that are forgotten afterwards. A typical response to the Learner's Paradox is a discussion of example, organization of observations, and analogy. Learning does often result in "new concept $\mathcal{B}$ is like old concept $\mathcal{A}$." Our efforts that follow do introduce serious examples at an early stage, and do return to them later as the concepts develop further. With Mauldin[1] and Ulam, we thus accept Shakespeare's advice:

All things done without example, in their issue
Are to be fear'd.

Several important results are demonstrated more than once but from different perspectives.

---

[1] Mauldin, R. D., Probability and nonlinear systems, *Los Alamos Science* (Special issue Stanislaw Ulam 1909–1984) **15**, 52 (1987).

Acknowledgements

We thank Dilip Asthagiri for a helpful reading of the whole manuscript. TLB acknowledges the National Science Foundation, the MURI program of the US Department of Defense, and David Stepp for support of this research. He also thanks Rob Coalson and Jimmie Doll for their many statistical mechanical insights. MEP thanks Themis Lazaridis, Rajesh Khare, Shekhar Garde, and Hank Ashbaugh for years of shared educational experiences in statistical thermodynamics, and Sam, Erin, and Annie for their unqualified understanding and encouragement throughout the years. LRP thanks Kathy, Jane, and Julia for their sympathy with a life of the mind.

Glossary

α_p	Coefficient of thermal expansion at constant pressure, $\alpha_p = -(1/\mathcal{V})(\partial \mathcal{V}/\partial T)_p$.
β	Inverse temperature in energy units, $\beta = 1/kT$.
γ_α	Activity coefficient for species α.
ΔU_α	Binding energy of a molecule of type α to the solution.
$\Delta \tilde{U}_\alpha$	Reference system contribution, indicated by the tilde, to the binding energy of a distinguished molecule of type α to the solution.
ϵ	Dielectric constant.
κ_T	The isothermal compressibility $\kappa_T = (-1/\mathcal{V})(\partial \mathcal{V}/\partial p)_T$.
$\bar{\kappa}_{T\alpha}$	The isothermal compressibility for pure liquid α.
Λ_α	Thermal de Broglie wavelength, $\Lambda_\alpha = \sqrt{\frac{h^2 \beta}{2\pi m_\alpha}}$, with m_α the mass of a molecule of type α.
μ_α	Chemical potential for molecular component α.
μ_α^{ex}	Interaction contribution to the chemical potential for molecular component α.
$\tilde{\mu}_\alpha^{\text{ex}}$	Interaction contribution to the chemical potential of species α due to reference interactions between a distinguished molecule and the solution.
ξ_α	Available volume fraction for species α.
ρ_α	Number density of species α.
$\bar{\rho}_\alpha$	Number density of species α in pure liquid α.
$\rho(\mathcal{N}, \mathcal{N}')$	Thermal density matrix.
ϕ_α	Volume fraction of species α in Flory–Huggins treatment of polymer mixtures.
$\chi_{\alpha\gamma}$	Flory–Huggins interaction parameter for mixing of pure liquids of species α and γ.

$\mathcal{X}_{\alpha\gamma}$ Interaction parameter in Flory–Huggins treatment of polymer mixtures; after normalization on a per monomer basis this becomes $\chi_{\alpha\gamma}$, also susceptibilities in discussion of density functional theories.

$\varphi_\alpha(\mathcal{R}^n)$ Applied external field acting on species α in conformation $\mathcal{R}^n$.

$\varphi_\alpha(r)$ Applied external field as a function of position acting on species α.

$a_{\alpha\gamma}^{(2)}$ Contribution of attractive interactions to the second virial coefficient for species pair $\alpha\gamma$; also van der Waals coefficient.

$\mathcal{A}$ Helmholtz free energy, $\mathcal{A} = \mathcal{E} - T\mathcal{S}$.

$b_\alpha(j)$ Indicator function; equal to one (1) when solvent molecule j occupies the inner shell of a distinguished molecule of type α, and zero (0) when this solvent molecule is outside that region.

$b_\alpha(\mathcal{R}^n)$ Indicator function; equal to one (1) for configurations $\mathcal{R}^n$ recognized as forming a molecule of type α, and zero (0) otherwise.

$b_{\alpha\gamma}^{(2)}(T)$ Second virial coefficient for species pair $\alpha\gamma$, as a function of temperature T, e.g. $b_{\alpha\gamma}^{(2)}(T) = \tilde{b}_{\alpha\gamma}^{(2)} - a_{\alpha\gamma}^{(2)}/kT$.

$\tilde{b}_{\alpha\gamma}^{(2)}$ Second virial coefficient for species pair $\alpha\gamma$, for the case of reference interactions only.

B_k Binomial moments, $B_k \equiv \left\langle \binom{n}{k} \right\rangle_0$.

c_n Concentration of a complex consisting of a specific solute of interest and n water molecules.

ε Binding energy for a distinguished molecule to the solution.

$\mathcal{E}$ Internal energy.

$\mathcal{G}$ Gibbs free energy, $\mathcal{G} = \mathcal{H} - T\mathcal{S}$.

$\hbar$ Planck constant divided by 2π, $\hbar = h/2\pi$.

$\mathcal{H}$ Enthalpy, $\mathcal{H} = \mathcal{E} + p\mathcal{V}$.

$\mathbb{H}$ Hamiltonian, $\mathbb{H} = \mathbb{K} + \mathbb{V}$.

k Boltzmann constant.

k_α Henry's Law coefficient for species α.

$\mathbb{K}$ Kinetic energy.

K_n Chemical equilibrium constant for the formation of a cluster of n water molecules with a specific solute of interest.

$m(x)$ For packing problems, the excess chemical potential as a function of available volume x, e.g. $m(x) = -\ln(1-x)$ is the primordial available volume theory.

M_α Empirical polymerization index for species α.

$\boldsymbol{n!}$ Boltzmann–Gibbs total number of permutations of identical molecules, $\boldsymbol{n!} \equiv \prod_\alpha n_\alpha!$.

p Pressure.

$\mathcal{P}_\alpha^{(0)}(\varepsilon)$ Probability distribution function for the binding energy of a distinguished molecule of type α to the solution in the case that these two subsystems are decoupled, as indicated by the superscript zero.

$\mathcal{P}_\alpha(\varepsilon)$	Probability distribution function for the binding energy for a distinguished molecule of type α to the solution in the case that these two subsystems are fully coupled.
$\tilde{P}_\alpha(\varepsilon)$	Distribution of the perturbative contribution to the binding energy of a distinguished molecule of type α to the solution in the case, indicated by the tilde, that the binding energy is solely derived from the reference system $\Delta \tilde{U}_\alpha$.
q_α^{int}	Internal partition function of a molecule of species α.
$\mathcal{Q}(\boldsymbol{n})$	Canonical partition function with volume $\mathcal{V}$ and temperature T not specifically indicated, $\mathcal{Q}(\boldsymbol{n}) = \mathcal{Q}(\boldsymbol{n}, \mathcal{V}, T)$.
$\mathcal{Q}(n_\alpha = 1)$	Canonical partition function for the case of exactly one molecule of type α, with volume $\mathcal{V}$ and temperature T (not specifically indicated), $\mathcal{Q}(n_\alpha = 1) = \mathcal{V} q_\alpha^{int} / \Lambda_\alpha^3$.
$\mathcal{R}^n$	Conformational coordinates for a molecule of n atoms.
$s_\alpha^{(0)}(\mathcal{R}^n)$	Distribution of conformational coordinates $\mathcal{R}^n$ for a molecule of type α with n atoms; in isolation as indicated by the superscript zero.
$\mathcal{S}$	Entropy.
T	Thermodynamic temperature.
$\bar{U}(\mathcal{N})$	Model potential energy function for incorporation of quantum mechanical effects in classical-limit configurational integral expressions for the canonical partition function.
$\bar{v}_\alpha$	Partial molar volume of pure liquid α, $\bar{v}_\alpha = 1/\bar{\rho}_\alpha$.
$\mathcal{V}$	Volume.
$\mathbb{V}$	Potential energy.
x_α	Mole fraction of species α.
z_α	Absolute activity of species α, $z_\alpha = e^{\beta \mu_\alpha}$.
$\boldsymbol{z}^{\boldsymbol{n}}$	Product of absolute activities, $\boldsymbol{z}^{\boldsymbol{n}} \equiv \prod_\alpha z_\alpha^{n_\alpha}$.
$\langle A|B \rangle$	Conditional expectation, the mean of A conditional on B.
$\langle\langle \ldots \rangle\rangle_0$	Expectation of '$\ldots$' under the circumstances of no coupling, indicated by the subscript zero (0), between solution and distinguished molecule.
$\langle \ldots | \mathcal{R}^n \rangle_0$	Expectation of '$\ldots$' conditional on the conformation $\mathcal{R}^n$ of the distinguished molecule.

1

Introduction

We consider a molecular description of solutions of one or more molecular components. An essential feature will be the complication of treating molecular species of practical interest since those chemical features are typically a dominating limitation of current work. Thus, liquids of atomic species only, and the conventional simple liquids, will only be relevant to the extent that they teach about molecular solutions. In this chapter, we will introduce examples of current theoretical, simulation, and experimental interest in order to give a feeling for the scope of the activity to be taken up.

The Potential Distribution Theorem (PDT) (Widom, 1963), sometimes called Widom's *particle insertion formula* (Valleau and Torrie, 1977), is emerging as a central organizing principle in the theory and realistic modeling of molecular solutions. This point is not broadly recognized, and there are a couple of reasons for that lack of recognition. One reason is that results have accumulated over several decades, and haven't been brought together in a unified presentation that makes that central position clear. Another reason is that the PDT has been primarily considered from the point of view of simulation rather than molecular theory. An initial view was that the PDT does not change simulation problems (Valleau and Torrie, 1977). In a later view, the PDT does assist simulations (Frenkel and Smit, 2002). More importantly though, it does give vital theoretical insight into molecular modeling tackled either with simulation or other computational tools, or for theory generally. This theorem has recently led to a new stage of molecular modeling of solutions, *quasi-chemical* theories that promise accurate molecular and chemical detail on the basis of available electronic structure computational methods of molecular science.

Our perspective is that the PDT should be recognized as directly analogous to the partition functions which express the Gibbsian ensemble formulation of statistical mechanics. From this perspective, the PDT is a formula for a thermodynamic potential in terms of a partition function. Merely the identification of a partition function does not *solve* statistical thermodynamics. But common ideas for approximate evaluation of partition functions can be carried over to the PDT. In contrast to Gibbsian partition functions, however, the PDT is built upon a local view of the thermodynamics and depends on local information. Therefore, the tricks from the common tool-kit work out simply and convincingly for liquids from the perspective of the PDT. In addition, the PDT gives a simpler perspective into some of the more esoteric ("... both difficult and strongly established ..." [Friedman and Dale, 1977]) results of statistical thermodynamics of molecular solutions. Thus, we present a point of view on the subject of molecular solutions that is simple, effective, and not developed elsewhere.

What follows is not a review. Nevertheless, there are old views underneath, and we do want to give the reader a valid sense of the scope, even historical scope, of the field.

Historical sketch

A genuine history is not offered here, but some historical perspective is required to appreciate what has been achieved. We suggest a natural division of that history into three periods: (a) a *pioneering* era prior to 1957 (the year that molecular simulation methods changed the field [Wood, 1986; Ciccotti *et al.*, 1987; Wood, 1996]), (b) the decade or so after 1957 when the theory of serious prototype liquid models achieved an impressive maturity, and (c) the present era including the past three decades, approximately.

Pioneering

The era before molecular simulation methods were invented and widely disseminated was a period of foundational scholarly activity. The work of that period serves as a basic source of concepts in the research of the present. Nevertheless, subsequent simulation work again revealed advantages of molecular resolution for developing detailed theories of these complex systems.

Students of this subject will remember being struck by the opinion expressed in the first English edition of the influential textbook *Statistical Physics* (Landau and Lifshitz, 1969), associated with this pioneering period:

"We have not included in this book the various theories of ordinary liquids and of strong solutions, which to us appear neither convincing nor useful."

A sensitive evaluation of the truth content of this view – it would not be a general view (Kipnis *et al.*, 1996) – is less important than the historical fact that it should be flatly asserted in that setting.

1957

The story of the initial steps in the development of molecular simulation methods for the study of liquids at the molecular scale has been charmingly recounted by Wood (1986; 1996). By 1957, these simulation techniques had been firmly established; the successful cross-checking of molecular dynamics calculations against Monte Carlo results was a crucial step in validating simulation methods for the broader scientific community.

Lots of ideas, many suboptimal, coexisted prior to the availability of the clear data that simulations provided. It was less that the simulations suggested new ideas than that the new simulation data served to alleviate the confusion of unclear ideas and to focus effort on the fruitful approaches. The theory of simple liquids treated by those simulations promptly made progress that we recognize, from our historical vantage point, as permanent. For example, the Percus–Yevick theory, proposed in 1957, was solved analytically for the hard-sphere case in 1963 (Wertheim, 1963; Thiele, 1963).

The PDT that is a central feature of this book dates from this period (Widom, 1963; Jackson and Klein, 1964), as does the related but separately developed scaled-particle theory (Reiss *et al.*, 1959).[1] Both the PDT and scaled-particle approaches have been somewhat bypassed as features of molecular theory, in contrast to their evident utility in simulation and engineering applications. Scaled-particle theories have been helpful in the development of sophisticated solution models (Ashbaugh and Pratt, 2004). Yet the scaled-particle results have been almost orthogonal to pedagogical presentations of the theory of liquids. This may be due to the specialization of the presentations of scaled-particle theory (Barrat and Hansen, 2003).

The theory of simple liquids wasn't simple

The theory of simple liquids achieved a mature state in the era 1965–1975 (Barker and Henderson, 1976; Hansen and McDonald, 1976). As this mature theory was extended towards molecular liquids, simple molecular cases such as liquid N_2 or liquid CCl_4 were treated first. But the molecular liquids that were brought within the perimeter of the successful theory were remote extremities compared with the liquids synthesized, poured from bottles or pipes, and used. In addition, the results traditionally sought from molecular theories (Rowlinson and Swinton, 1982) often appear to have shifted to accommodate the limitations of the available theories. Overlooking molecular simulation techniques

---

[1] Kirkwood and Poirier (1954) had earlier used a result equivalent to the PDT in a specialized context, and Stell (1985) discusses Boltzmann's use of an equivalent approach for the hard-sphere gas.

for the moment, it is difficult to avoid the conclusion that methods at the core of the theory of simple liquids stalled in treating molecular liquids of interest. A corollary of this argument is that molecular simulations overwhelmingly dominate theoretical activity in the theory of molecular liquids. The RISM (reference interaction site method) theory, then (Chandler and Andersen, 1972) and now (Hirata, 2003) an extension of successful theories of simple liquids, is an important exception. The luxuriant development of RISM theories highlights aspects of the theory of simple liquids that were left unsolved, particularly the lack of a "... theory of theories ..." in the memorable words of Andersen (1975). Similarly, the introduction of the *central force models* for liquid water (Lemberg and Stillinger, 1975) broached the practical compromise of treating molecular liquids explicitly as complicated atomic liquids. A primary reason for those new models was that the theory of molecular liquids seemed stuck at places where the theory of atomic liquids was soundly developed. Despite the novelty of the central force models, the eventual definitive theoretical study (Thuraisingham and Friedman, 1983) of the central force model for water using the classic hypernetted chain approximation of the theory of simple liquids was disappointing, and highlighted again the lack of a theory of those theories of atomic liquids as they might be relevant to theories of molecular liquids.

The category of *simple* liquids is sometimes used to establish the complementary category of *complex* liquids (Barrat and Hansen, 2003). Another and a broad view of complex liquids is that they are colloid, polymer, and liquid crystalline solutions featuring a wide range of spatial length scales – sometimes called soft matter (de Gennes, 1992). Planting ourselves at an *atomic* spatial resolution, the models analyzed for those complex liquids are typically less detailed and less realistic on an atomic scale than models of atomic liquids.

In this book, simple liquids are contrasted with molecular liquids. Our goal is to treat molecular liquids on an atomic spatial scale. That doesn't mean approximations, perhaps even crude ones, won't sometimes be considered. But theories of molecular liquids, even with appropriate approximations, require molecule-specific features which the theory of simple liquids doesn't supply. As an example, the molecular liquid water is acknowledged here as a non-simple liquid, and a particularly complex molecular liquid.

1.1 Molecules in solution

Here we give a series of examples that illustrate our interest in molecules in solutions, and exemplify the problems that motivate the theory and modeling that is the subject of the remainder of this book. Many of the issues that are discussed physically here will be studied in detail in subsequent chapters.

Reversed-phase liquid chromatography

Reversed-phase liquid chromatography (RPLC) is a workhorse technique for the separation of water-soluble chemical species. The method is used for chemical separations of molecular mixtures with sizes ranging from small molecules to biomolecules. Small differences in free energies, on the order of $0.1 \, \text{kcal mol}^{-1}$ for transfer between the aqueous mobile phase and the stationary phase of tethered alkanes, can lead to well-resolved chromatographic peaks following passage through the column. Understanding the driving forces for retention is thus a severe challenge to the theory of molecular liquids.

A typical RPLC column is packed with porous silica beads. The pore sizes have a length scale in the range 50–100 Å. The silica substrate is derivatized by a silanization reaction which attaches alkanes to hydroxyl groups on the surface. Chains of length 8–18 carbon atoms are utilized, with the C_{18} case being common. Surface densities up to about 50% of close packing are typical. The mobile phase most often contains a cosolvent, such as methanol or acetonitrile, in addition to water.

The flow rate is extremely slow on the time scale of molecular motions. Therefore, a quasi-equilibrium treatment is valid. The retention factor – the difference between the retention time for a peak of interest and a standard unretained reference, divided by the retention time for the reference – can then be taken as

$$k' = \Phi K, \tag{1.1}$$

where k' is the retention factor, Φ is the ratio of volumes of the stationary and mobile phases, and K is an equilibrium constant for solute partitioning between the two phases. The equilibrium constant is

$$K = \exp\left[-\beta \Delta \mu^{\text{ex}}\right], \tag{1.2}$$

involving the excess chemical potential difference for the solute between the mobile and stationary phases. Thus, equilibrium statistical mechanics provides a framework to examine the free energy driving forces for retention.

Extensive experimental, theoretical, and modeling work has been directed at revealing details of the complex interface between the mobile and stationary phases (Beck and Klatte, 2000; Slusher and Mountain, 1999; Zhang *et al.*, 2005). Current debates are centered on the issue whether retention can be correctly understood as bulk liquid-phase partitioning or an adsorption process. The simulation work has shown that it is really neither of these two extremes; the interface possesses specific ordering features that present a non-bulk environment, yet nonpolar solutes do penetrate significantly into the stationary phase. Also, the measured excess chemical potential change is typically in line with bulk partitioning values.

Figure 1.1 shows a snapshot of the interface between a C_{18} stationary phase (at 50% coverage) and a 90/10, by volume, water/methanol mobile phase. Clearly the stationary phase is disordered, with little penetration of the aqueous phase into the stationary phase.

The density profiles for the 90/10 mobile-phase case of Fig. 1.1 are shown in Fig. 1.2. Notice the layering of the alkane chain segments near the solid surface. The chains topple over and pack into a disordered layer on the surface in order to fill the available volume. The chain segments never attain a flat density profile, so this region cannot be considered a bulk alkane fluid. In addition, the first $10\,\text{Å}$ of the stationary phase near the silica support is glassy, while the tails exhibit liquid-like diffusive motion. The width of the stationary phase is about $18\,\text{Å}$, smaller than the fully extended length of $24\,\text{Å}$. The aqueous mobile phase does not penetrate significantly into the stationary phase, which is expected for aqueous/hydrophobic interfaces. Notice also the build-up of methanol at the interface, consistent with the observed reduction in surface tension in water/methanol mixtures. Previous simulations (Beck and Klatte, 2000) have shown that, beyond the segregation to the surface, methanol molecules in the interfacial region preferentially point their nonpolar ends toward the stationary phase. The waters are not orientationally ordered at the interface to an appreciable extent.

The excess chemical potential profile for methane gives a first indication of the driving force for retention (Fig. 1.3). This profile was computed using the PDT discussed in this book. A methane molecule experiences a $2.5\,\text{kcal}\,\text{mol}^{-1}$ free energy drop on passing into the stationary phase, consistent with a bulk partitioning value; the drop on the right is at the liquid/vapor interface.

The free energy contributions to the transfer energy can be separated into repulsive and attractive components by examining the excess chemical potential profile for hard-sphere solutes (Fig. 1.4). For the pure aqueous mobile phase case in Fig. 1.4, there is a clear free energy minimum at the interface, indicating an increase in available volume because of the weak interpenetration of the hydrophobic and aqueous components. Thus the purely repulsive contribution to the free energy drives the solute into the interfacial region, while the attractive component leads to an additional drop entering the interface and further penetration into the stationary phase due to interactions with the tethered alkanes. With an increase of methanol content, the excess free volume for the hard-sphere solute disappears, suggesting better penetration of methanol into the stationary phase. A lesson from these simulation studies is that molecular-level realism is required to tackle these complicated interfacial problems.

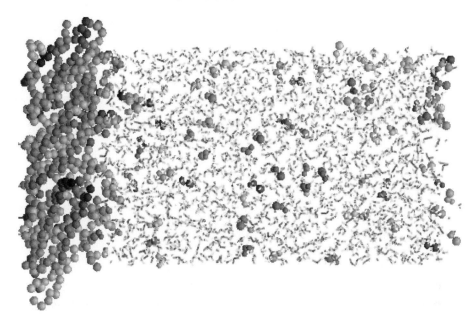

Figure 1.1 Snapshot of a reversed-phase liquid chromatographic interface. Tethered C_{18} chains are leftmost, displayed as ball-and-stick. The water/methanol mixture mobile phase, 10% methanol by volume, is on the right terminated by a fluid interface with vapor. Three-site model methanol molecules are also displayed as ball-and-stick, but the water molecules are wires only. (See Clohecy, 2005.)

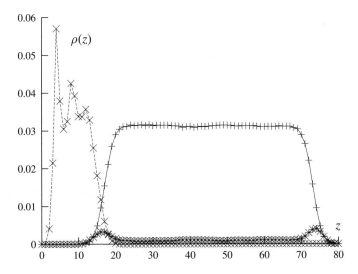

Figure 1.2 Number density profiles for the C_{18} carbon atoms, water oxygen atoms, and methanol carbon atoms for the 10% methanol case. The alkane carbon density is on the left, and the high-density profile is for water oxygen on the right. Distances along z perpendicular to the interfacial plane are given in Å. Densities are given in arbitrary units.

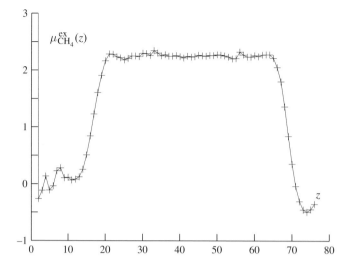

Figure 1.3 Excess chemical potential profile for methane for the 90/10 water/methanol mobile phase. Energies are given in $kcal\,mol^{-1}$. Distances along z perpendicular to the interfacial plane are given in Å.

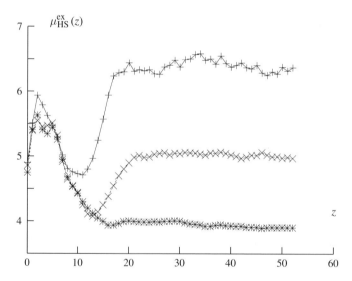

Figure 1.4 Excess chemical potential profiles ($kcal\,mol^{-1}$) for a 2.0 Å hard-sphere solute along the z axis. The C_{18} tethered chains are on the left. The three curves correspond to pure water (top), a 50/50 mixture (middle), and 90% methanol (bottom) by volume. Energies are in $kcal\,mol^{-1}$. Distances along z perpendicular to the interfacial plane are given in Å.

1.1 Compare expected free energy driving forces for retention of nonpolar *vs.* polar solutes in relation to the specific interfacial ordering effects in the RPLC system discussed above.

Variation of the dissociation constant of triflic acid with hydration

Nafion® is an archetypal proton-conducting material used in fuel-cell membranes. It is a Teflon®-based polymeric material with side chains ending with a hydrophilic sulfonic acid group. Because of these super-acidic head groups, Nafion® can be hydrated and can serve as a proton conductor separating electro-chemical compartments. The consequences of variable hydration on proton conduction is expected to be significant for the performance of these materials. Experimental work has not yet fully resolved these issues, and a molecular understanding is still sought.

The important acid activity in Nafion® is appropriately represented by trifluoromethane sulfonic (triflic) acid, CF_3SO_3H; see Fig. 1.5. Dielectric spectroscopy has suggested that a significant amount of triflic acid is not dissociated in the ionic melt at 50% mole fraction of water (Barthel *et al.*, 1998). But the deprotonation chemistry of hydrated triflic acid hasn't been experimentally studied over the wide range of hydration and temperature that would be relevant to the function of sulfonate-based polyelectrolyte membrane materials.

Here we discuss a simple theoretical molecular model of triflic acid dissociation in dilute aqueous solution along the gas–liquid saturation curve to elevated

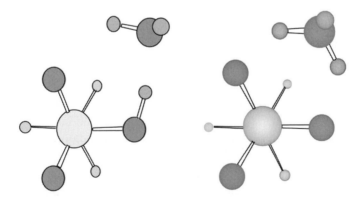

Figure 1.5 A trifluoromethane sulfonic (triflic) acid molecule, CF_3SO_3H, with one water molecule. The picture on the left shows the approximate minimum energy configuration of these molecules. The view is down the SC bond, with the SO_3 group closest to the viewer and the CF_3 behind. The picture on the right shows a configuration proposed as a transition state for scrambling of the acid hydrogen. (See Paddison *et al.*, 1998.)

temperatures. The water liquid–vapor saturation curve serves as a simple repro-
ducible path for reduction of the hydration of the triflic acid. These results are
preparatory to molecular modeling at higher resolution (Eikerling *et al.*, 2003).
Since data that would make this effort more conclusive are not yet available, these
theoretical considerations might also serve to encourage further experimental work
in this direction.

The approach here will be to take a specific triflic acid molecule, a *distinguished*
molecule, as the basis for detailed molecular calculations, but to idealize the
solution external to that molecule as a dielectric continuum. "External" will be
defined as outside a molecular cavity established with spheres on each atom,
characterized by radii treated as parameters. We will follow standard procedures
in applying this dielectric model (Tawa and Pratt, 1994; 1995; Corcelli *et al.*,
1995). The specific goal below will be to treat the equilibrium

$$HA + H_2O \rightleftharpoons A^- + H_3O^+, \tag{1.3}$$

with $A^- = CF_3SO_3^-$, over an extended range of conditions. The equilibrium ratio

$$K_a = \frac{\rho_{A^-}\rho_{H_3O^+}}{\rho_{HA}\rho_{H_2O}} \tag{1.4}$$

can be expressed as

$$K_a = K_a^{(0)} \exp\left[-\beta \Delta\mu^{ex}\right], \tag{1.5}$$

with

$$\Delta\mu^{ex} \equiv \mu_{A^-}{}^{ex} + \mu_{H_3O^+}{}^{ex} - \mu_{HA}{}^{ex} - \mu_{H_2O}{}^{ex}, \tag{1.6}$$

and μ_α^{ex} is the interaction contribution to the chemical potential of species α. The
modeling of each μ_α^{ex} is the principal topic of this book. $\rho_\alpha = n_\alpha/\mathcal{V}$ is the number
density or concentration of species α. The factor $K_a^{(0)}$ is the equilibrium ratio found
from the molecular computational results for the reactions without consideration
of a solution medium; each of the interaction contributions of Eq. (1.6) may be
estimated with the dielectric model. We will be interested here in conditions of
infinite dilution of the solute but a wide range of conditions for the solvent;
thus ρ_{H_2O} will vary widely. This warrants the appearance of the water density in
Eq. (1.4), which is typically omitted when the water density has a standard value
only.

Consideration of the isodesmic (Hehre *et al.*, 1970; McNaught and Wilkinson,
1997) Eq. (1.3) carries assumptions about the chemical state of the dissociated
proton. Here our primary interest is the triflic acid, however, so we assume that this
treatment is satisfactory for estimating hydration free energies without consideration of the involved issues of H^+(aq). If those assumptions were uncomfortable,

and if the $OH^-(aq)$ were viewed as chemically simpler, then we could study the isodesmic equilibrium

$$HA + OH^- \rightleftharpoons A^- + H_2O, \qquad (1.7)$$

with equilibrium ratio

$$K' = \frac{\rho_{A^-}\rho_{H_2O}}{\rho_{HA}\rho_{OH^-}}. \qquad (1.8)$$

Then

$$K_a = \frac{K'K_W}{\rho_{H_2O}^2}, \qquad (1.9)$$

with the water ion product $K_W = \rho_{H_3O^+}\rho_{OH^-}$ taken to have its empirically known value (Tawa and Pratt, 1994).

Dielectric hydration models serve as primitive theories against which more detailed molecular descriptions can be considered. Of particular interest are temperature and pressure variations of the hydration free energies, and this is specifically true also of hydrated polymer electrolyte membranes. The temperature and pressure variations of the free energies implied by dielectric models have been less well tested than the free energies close to standard conditions. Those temperature and pressure derivatives would give critical tests of this model (Pratt and Rempe, 1999; Tawa and Pratt, 1994). But we don't pursue those tests here because the straightforward evaluation of temperature and pressure derivatives should involve temperature and pressure variation of the assumed cavity radii about which we have little direct information (Pratt and Rempe, 1999; Tawa and Pratt, 1994).

To frame this point, we give simple estimates of temperature and pressure derivatives assuming that the thermodynamic state dependence of the radii may be neglected. We will consider a simple ion and the Born formula (Pettitt, 2000); the interaction contribution to the chemical potential of such a solute is $\mu^{ex} \approx -\frac{q^2}{2R}\frac{(\epsilon-1)}{\epsilon}$. Here q is the charge on the ion and R is its Born radius; see Section 4.2. We assume that these radius parameters are independent of the thermodynamic state. Considering the partial molar volume first, we have

$$v^{ex} \equiv \left(\frac{\partial\mu^{ex}}{\partial p}\right)_{T,n} \approx -\frac{q^2}{2R\epsilon}\left(\frac{\partial\ln\epsilon}{\partial p}\right)_T = -\frac{q^2}{2R\epsilon}\left(\frac{\partial\ln\epsilon}{\partial\ln\rho}\right)_T \kappa_T. \qquad (1.10)$$

The superscript ^{ex} indicates that this is the contribution due to solute–solvent interactions; it is the contribution in excess of the ideal gas at the same density and temperature. Our later developments will make it clear how this contributes to the full chemical potential and the related standard state issues (Friedman and Krishnan, 1973). $\kappa_T = (-1/\mathcal{V})(\partial\mathcal{V}/\partial p)_T$ is the isothermal coefficient of bulk compressibility of the pure solvent. The required density derivative of the dielectric constant is evaluated with the fit of Uematsu and Franck (1980), yielding $(\partial \ln \epsilon / \partial \ln \rho)_T \approx 1.15$ at the standard point $T = 298.15\,\mathrm{K}$ and $\rho = 997.02\,\mathrm{kg\,m^{-3}}$. We further estimate $\kappa_T \approx 46 \times 10^{-6}\,\mathrm{atm^{-1}}$ (Eisenberg and Kauzmann, 1969, see Figs. 4.14 and 4.15). Finally, the leading factor $q^2/2R\epsilon$ is of the order of $1\,\mathrm{kcal\,mol^{-1}}$ or about $40\,\mathrm{cm^3\,atm\,mol^{-1}}$. The combination Eq. (1.10) thus gives us an order of magnitude of $2 \times 10^{-3}\,\mathrm{cm^3\,mol^{-1}}$. Experimental results are typically a thousandfold larger.

For the temperature dependence we estimate

$$s^{\mathrm{ex}} \equiv -\left(\frac{\partial \mu^{\mathrm{ex}}}{\partial T}\right)_{p,n} \approx \frac{q^2}{2R\epsilon}\left[\left(\frac{\partial \ln \epsilon}{\partial T}\right)_\rho - \left(\frac{\partial \ln \epsilon}{\partial \ln \rho}\right)_T \alpha_p\right]. \tag{1.11}$$

$\alpha_p = -(1/V)(\partial V/\partial T)_p$ is the coefficient of thermal expansion for the pure solvent. The additional temperature derivative is $(\partial \ln \epsilon / \partial T)_\rho \approx -4.3 \times 10^{-3}\,\mathrm{K^{-1}}$ (Uematsu and Franck, 1980), at the standard point indicated above, and $\alpha_p \approx 3 \times 10^{-4}\,\mathrm{K^{-1}}$ (Eisenberg and Kauzmann, 1969). This entropy contribution is negative and has a magnitude of a small multiple of $1\,\mathrm{cal\,K^{-1}\,mol^{-1}}$. This magnitude is about a power of ten smaller than typical experimental results. Again, notice that this doesn't make a comparison that would warrant detailed discussion of a standard state for a particular experiment (Friedman and Krishnan, 1973).

Despite the neglect here of the variation of the radius parameter R with thermodynamic state, this model does address the availability of water at a primitive level through Eq. (1.3). In addition, it is a physical model – in language to be adopted later, this is a *realizable* model – and it is sufficiently simple to be helpful.

The variation of the defined dissociation constant, obtained on the basis of this dielectric model, is plotted in Fig. 1.6. The reaction Eq. (1.3) in liquid water becomes unfavorable from the perspective of the free energy upon exceeding $500\,\mathrm{K}$ on the saturation curve, where the liquid density falls below about 85% of the triple-point density. Nevertheless, this sulfonic acid head group would still be considered a strong acid in bulk aqueous solution at these elevated temperature and reduced-density conditions. These results give perspective for the view that insufficient hydration can result in incomplete dissociation of sulfonic acid species in membranes.

The static dielectric constant of liquid water is roughly 30 in this interesting region around $500\,\mathrm{K}$ on the liquid–vapor coexistence curve. If a static dielectric constant were assigned to a hydrated Nafion® membrane matrix, the value is unlikely to be significantly larger than 30. In this respect, the present model is not extreme. But heterogeneity of the local environment of a sulfonic acid head group is probably a significant factor in both of the physical systems, liquid water and hydrated Nafion® membranes, considered. The physical picture of a dielectric continuum model is more uniform than either of those systems. In this respect, this model calculation probably gives a limiting possibility of the effect

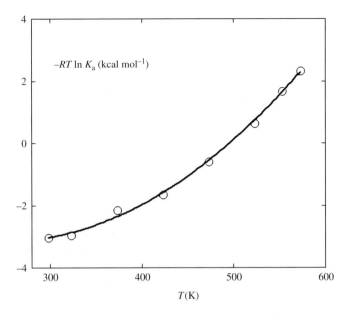

Figure 1.6 Variation of the dissociation coefficient Eq. (1.5) of reaction Eq. (1.3) with temperature along the water saturation curve, as described by a dielectric continuum model. $R = 1.987\,\text{cal mol}^{-1}\,\text{K}^{-1}$ here is the gas constant. See Paddison *et al.* (2001).

of hydration on the degree of dissociation of Nafion® head groups. The model employed here is reasonable but not compelling. To test and refine the present conclusions, further experiment is necessary. A better developed molecular theory would also be useful. That is the topic of this book.

Exercises

1.2 Show that Eq. (1.5) is an exact expression for the acid dissociation constant. Developments in Chapter 3 will make this problem transparent.

1.3 Work out Eqs. (1.10) and (1.11).

Ion channels

As a final example, we discuss biological ion channels (Hille, 2001). Ion channels are protein structures embedded in biological membranes. They control the flux of ions into and out of the cell under various stimuli, including voltage, pH, and ligand binding. Ion channels allow for the passive diffusion of the ions under concentration or electrostatic potential gradients. They are also implicated in a wide range of genetic disorders (Ashcroft, 2000). Two key features of ion channels are their selectivity and gating, opening or closing of the channel due to a stimulus. Selectivity is based on details of intermolecular interactions, including ion size, electrostatic interactions with the protein, and hydration. And gating the pore can involve either large-scale motions of the protein, or movement of local groups. The challenge to theory and modeling is to establish a valid molecular-level understanding of these important functions.

In the last several years, X-ray structures of several ion channels have appeared. These structures include bacterial K^+ and Cl^- channel homologs (MacKinnon, 2003; Dutzler *et al.*, 2002; 2003). The structures themselves give insights into how the channels operate. The X-ray structure is only a snapshot, however, and molecular theory and simulation can add to a more detailed picture of the channel function. Figure 1.7 shows a ribbon diagram of a bacterial Cl^- channel homolog. This structure consists of two identical monomer units, and since each unit contains a pore for Cl^- conduction, Fig. 1.7 shows a double-barrelled pore. The ion conduction path is not a straight tunnel through the protein (Yin *et al.*, 2004),

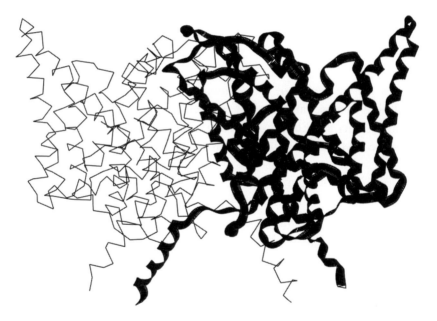

Figure 1.7 Bacterial Cl^- channel homolog structure. The dimeric protein is viewed from the side (the plane of the image is perpendicular to the membrane plane). The two monomers are separated by a line down the middle of the figure. The membrane is omitted. Each monomer contains a separate pore for Cl^- ion permeation. Each pore contains a selectivity filter and a gating domain which is believed to involve a pH-sensitive glutamate residue. The location of the ion transit pathways and the mechanism of transport through the pores are topics of current channel research (Yin *et al.*, 2004).

as for the K^+ channel.[2] Conductivities in eukaryotic chloride channels have been studied for an array of ions, including Br^-, NO_3^-, SCN^-, and ClO_4^- in addition to Cl^- (Maduke and Miller, 2000). Work is currently in progress on several aspects of the ion channels and transporters: (1) where is the conduction path? (2) what is the origin of selectivity? (3) how does the channel gate? (4) what are key ligand or drug binding sites on the intracellular and extracellular sides? (5) how does pH affect channel properties?

One topic in the study of channels that raises questions highly relevant to the subject of this book is the nature of the binding of channel blockers to the pore. In the case of potassium channels, a common blocking agent is the tetraethylammonium (TEA) molecular ion (Crouzy *et al.*, 2001; Thompson and Begenisich, 2003). Alkylammonium ions exhibit complex phase behavior in aqueous solution including indications of hydrophobic effects (Weingartner *et al.*, 1999). The nature of the binding to the pore is an interplay between ionic interactions and hydration phenomena (Crouzy *et al.*, 2001).

Studying the fundamental issues in ion channels is a stimulating field connecting molecular theory to biophysics and biomedical research. The PDT can contribute in several ways to understanding basic aspects of channels. Membrane permeability can be described by

$$P = DK/L, \tag{1.12}$$

where P is the permeability, D is the ion diffusion constant in the membrane channel, K is a partition coefficient analogous to Eq. (1.2), and L is the width of the membrane (Aidley and Stanfield, 1996). Generally the predominant contribution to selectivity is the ion partition coefficient, and thus the excess chemical potential profile figures prominently in determining ion selectivity. Accurate treatment of electrostatic effects is required. Effects of ion size and protonation states of key residues must also be addressed with detailed molecular-level theory. These are daunting problems, since a simulation of the protein, membrane, and surrounding solvent typically involves 100 000 atoms or more, and the time scale for a single ion passage through the channel is on the order of 10–100 ns.

Exercise

1.4 The potassium channel can successfully discriminate between K^+ and Na^+ ions. It is interesting to note that the partial molar volume of $Na^+(aq)$ is negative, whereas the partial molar volume of K^+ is positive (see Table VIII

---

[2] To illustrate the complexities of ion channels, there is recent evidence that the bacterial ClC "channel" is actually an H^+/Cl^- transporter (Accardi and Miller, 2004). The bacterial and eukaryotic structures do possess strong similarity, however, and the eukaryotic case ClC–0 is indeed a true channel (Chen and Chen, 2001).

in Friedman and Krishnan, 1973). Examine the structure of the potassium channel presented in MacKinnon (2003), and assume that the major effect is due to the partition coefficient from bulk solution into the channel. What are the possibilities for the origin of the excess chemical potential differences that might lead to this discrimination?

1.2 Looking forward

Which real liquids present important research issues, and why? Here we sketch an idiosyncratic response to such questions (Pratt, 2003). This serves to clarify the emphases that this book will make. A rough estimate of rank, from more to less important, is:

a. *Water and aqueous solutions.* This ranking is due to the ubiquity of water in our physical world and our culture. Water has a high curiosity value because it is peculiar among liquids and often participates in chemical reactions. Molecular biophysics and nanotechnology are closely related because molecular biophysics offers numerous examples of sophisticated, molecular-scale mechanical, electrical, and optical machines. Water participates directly in the structure and function of these machines. Aqueous interfaces, particularly involving the structure and thermodynamics of surfactant layers, are intrinsic to molecular biology and modern technology.

b. *Petroleum-derived liquids.* This includes the vast majority of solvents used in chemical processes, most nonbiological polymeric solutions, and most nonbiological liquid crystal systems. Polymers, liquid crystals, surfactants, and nanotechnology overlap items (a) and (b) here. One view is that the chemical engineering profession, historically dominated by the science of petroleum fluids, will in the future significantly include bio- and nanotechnology problems (Lenhoff, 2003). This view reveals the perspective that chemical engineers are a sophisticated audience for good theories of liquids.

c. *Plasmas, the Earth's interior, the Sun, high-density reacting fluids.* In these cases "fluid" typically seems a more relevant appellation than "liquid." Fluid metals are typically legitimate liquids, and chemically complex *room temperature ionic liquids,* organic salts that melt at low temperatures, are of high current interest.

d. *Liquid helium, the gaseous giant planets, and quantum liquids.* Both items (c) and (d) have high curiosity values, but, with the noted exception of room temperature ionic liquids, do not present the complexities that distinguish molecular liquids.

e. *Simple liquids.* First and foremost, this has meant liquid argon. But other liquefied simple gases, such as Ne, N_2, O_2, and CH_4, occupy this category too. Work on this category is foundational: to establish that the simplest cases can, in fact, be well solved.

1.3 Notation and the theory of molecular liquids

Our principal interest in this book will be the molecular basis of the theory of molecular solutions. Because the molecular components may themselves be complicated, the notation can be complicated. This can be a huge, non-physical problem for this field.[3] We address this issue in a couple of ways that are noted here. First, we express important results in *coordinate-independent* ways. This mostly means expressing important results in terms of averages that might be obtained from a simulation record. Molecular simulation does require coordinate choices. But our strategy in this is consistent with the view that simulation calculations are the most accessible source of primitive data for those interested in the theoretical basis of these problems.

The potential distribution theorem, which is central to this book, lends itself to this goal of coordinate independence. But literal coordinate independence is not possible where we really want to know how molecules achieve interesting results. Thus we do need some notation for coordinates, and we use $\mathcal{R}^n$ generically to denote the configuration of a molecule of n atoms, including translational, orientational, and conformational positioning; see Fig. 1.8. This notation suggests that cartesian coordinates for each atom would be satisfactory, in principle, but doesn't require any specific choice. The notation $\mathcal{R}^n$ would be too cumbersome for indicating the configurations of several specific molecules together; we typically write (i, j) to indicate the configuration of the specific molecules i and j, and $\mathcal{N}$

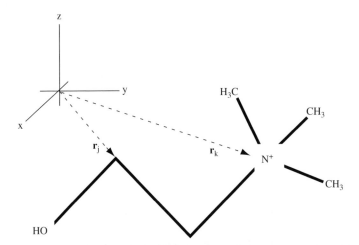

Figure 1.8 Illustration of the notational definition of conformational coordinates $\mathcal{R}^n = \{r_1, r_2, \ldots, r_n\}$. The conformational distribution $s_\alpha^{(0)}(\mathcal{R}^n)$ – see Eq. (1.13), p. 18 – is sampled for the single molecule in the absence of interactions with solvent by suitable simulation procedures using coordinates appropriate for those procedures. The normalization adopted in this development is $\int s_\alpha^{(0)}(\mathcal{R}^n)\mathrm{d}(\mathcal{R}^n) = \mathcal{V}$, the volume of the system. Further details for a common example can be found in Section 2.1, p. 23.

---

[3] One perspective is that graph theory methods of the equilibrium theory of classical liquids are merely simple, common, intuitive notations. It is then ironic that these notations are typically a *denouement* of an extended theoretical development (Uhlenbeck and Ford, 1963; Stell, 1964).

for the full configuration of the N specific molecules. A simple consistent extension is that $N + 1$ is the configuration of the N-plus-first specific molecule and $(\mathcal{N}, \mathcal{N} + 1)$ is the full configuration of an $N + 1$ molecule system. This notation, which has some precedent, will conflict somewhat with a standard thermodynamic notation where $\boldsymbol{n} = \{n_1, n_2, \ldots\}$ is a collection of particle numbers. The context should distinguish the meaning in those cases.

We expect that the typical initial response to the fundamental statistical mechanical formulae presented as in Eq. (2.15), p. 26, will be that this notation is schematic. We reply that for molecular liquids any communicative notation will be schematic. For example, cartesian positions of molecular centers is schematic, and for our purposes misses the important point of molecular liquids. We hope that the schematic notation that is employed will communicate satisfactorily.

Finally, we adopt a notation involving conditional averages to express several of the important results. This notation is standard in other fields (Resnick, 2001), not without precedent in statistical mechanics (Lebowitz *et al.*, 1967), and particularly useful here. The joint probability $P(A, B)$ of events A and B may be expressed as $P(A, B) = P(A|B)P(B)$ where $P(B)$ is the marginal distribution, and $P(A|B)$ is the distribution of A conditional on B, provided that $P(B) \neq 0$. The expectation of A conditional on B is $\langle A|B \rangle$, the expectation of A evaluated with the distribution $P(A|B)$ for specified B. In many texts (Resnick, 2001), that object is denoted as $E(A|B)$ but the bracket notation for "average" is firmly established in the present subject so we follow that precedent despite the widespread recognition of a notation $\langle A|B \rangle$ for a different object in quantum mechanics texts.

The initial introduction of conditional probabilities is typically associated with the description of *independent* events, $P(A, B) = P(A)P(B)$ when A and B are independent. Our description of the potential distribution theorem will hinge on consideration of independent systems: first, a specific distinguished molecule of the type of interest and, second, the solution of interest. We will use the notation $\langle\langle \ldots \rangle\rangle_0$ to indicate the evaluation of a mean, average, or expectation of "$\ldots$" for this case of these two independent systems. The doubling of the brackets is a reminder that two systems are considered, and the subscript zero is a reminder that these two systems are independent. Then a simple example of a conditional expectation can be given that uses the notation explained above and in Fig. 1.8:

$$\langle\langle \ldots \rangle\rangle_0 = \mathcal{V}^{-1} \int \langle \ldots | \mathcal{R}^n \rangle_0 \, s_\alpha^{(0)} (\mathcal{R}^n) \mathrm{d} \left(\mathcal{R}^n \right). \tag{1.13}$$

The second set of brackets isn't written on the right here because the second averaging is explicitly written out. This is understood by recognizing that the $\mathcal{V}^{-1} s_\alpha^{(0)} (\mathcal{R}^n)$ is the normalized thermal distribution of configurations of the distinguished molecule, and that the bracket notations indicate independence for the molecule and solution configurations.

We conclude this section by giving a topical example of the utility of conditional averages in considering molecularly complex systems (Ashbaugh *et al.*, 2004). We considered the RPLC system discussed above (p. 5), but without methanol: n-C_{18} alkyl chains, tethered to a planar support, with water as the mobile phase. The backside of the liquid water phase contacts a dilute water vapor truncated by a repulsive wall; see Fig. 1.2, p. 7. Thus, it is appropriate to characterize the system as consistent with aqueous liquid–vapor coexistence at low pressure. A standard CHARMM force-field model (MacKerell Jr. *et al.*, 1998) is used, as are standard molecular dynamics procedures – including periodic boundary conditions – to acquire the data considered here. Our interest is in the interface between the stationary alkyl and the mobile liquid water phases at 300 K.

This system displays (Fig. 1.9) a traditional interfacial oxygen density profile that has been the object of measurement (Pratt and Pohorille, 2002), monotonic with a width 2–3 times the molecular diameter of a water molecule. This width is somewhat larger than that of water–alkane liquid–liquid interfaces, though it is still not a broad interface. The enhanced width is probably associated with roughness of the stationary alkyl layer; the carbon density profile is shown in Fig. 1.9 as well.

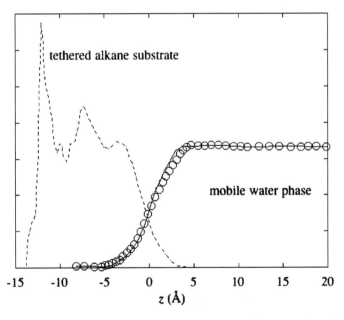

Figure 1.9 Carbon- and water-oxygen interfacial densities as a function of z. The dashed and solid lines indicate the observed carbon and oxygen densities, respectively, at 300 K determined from molecular simulation. The disks plot the water-oxygen densities reconstructed from the proximal radial distribution function for carbon-oxygen (see Fig. 1.2), averaged over alkyl chain conformations sampled by the molecular simulation. The interfacial mid-point ($z = 0$) is set at the point where the alkyl carbon- and water-oxygen densities are equal. See Figs. 1.1 and 1.2, p. 7.

Consider now the mean oxygen density *conditional* on a specific alkyl configuration. Since that conditional mean oxygen density is less traditionally analyzed than the density profile shown in Fig. 1.9, we exploit another characterization tool, the *proximal* radial distribution (Ashbaugh and Paulaitis, 2001). Consider the volume that is the union of the volumes of spheres of radius r centered on each carbon atom; see Fig. 1.10. The surface of that volume that is closer to atom i than to any other carbon atom has area $\Omega_i(r) r^2$ with $0 \leq \Omega_i(r) \leq 4\pi$. The proximal radial distribution function $g_{\text{prox}}(r)$ is defined as

$$\langle n_{\text{W}}(r; \Delta r)\rangle = \left(\sum_i \Omega_i(r)\right) r^2 \rho_{\text{W}} g_{\text{prox}}(r) \Delta r. \tag{1.14}$$

Here $\langle n_{\text{W}}(r; \Delta r)\rangle$ is the average number of atoms (oxygen or hydrogen) in the shell volume element, of width Δr, that tracks the surface of the volume described above, and ρ_{W} is the bulk water density. If the alkyl configuration were actually several identical atoms widely separated from one another, the $g_{\text{prox}}(r)$ for oxygen atoms defined in this way would be just the conventional radial distribution of oxygen conditional on one such atom. More generally, by treating the actual solid angles $\Omega_i(r)$ this formula attempts to clarify structural obfuscation due to blocking by other carbon units. The $\Omega_i(r)$ were calculated here by a Monte Carlo sampling; these properties of water densities conditional on chain conformations were then averaged over observed alkyl configurations.

The proximal radial distribution functions for carbon–oxygen and carbon–(water)hydrogen in the example are shown in Fig. 1.11. The proximal radial distribution function for carbon–oxygen is significantly more structured than the interfacial profile (Fig. 1.9), showing a maximum value of 2. This proximal radial distribution function agrees closely with the carbon–oxygen radial distribution function for methane in water, determined from simulation of a solitary methane molecule in water. While more structured than expected from the

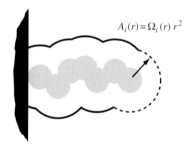

Figure 1.10 Geometrical quantities in defining the proximal radial distribution function $g_{\text{prox}}(r)$ of Eq. (1.14). The surface proximal to the outermost carbon (carbon i), with area $\Omega_i(r) r^2$, permits definition of the mean oxygen density in the surface volume element, conditional on the chain configuration: $\rho_{\text{W}} g_{\text{prox}}(r)$.

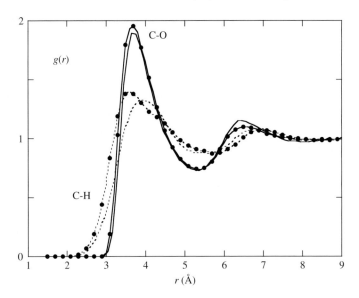

Figure 1.11 Carbon–water proximal and radial distribution functions at 300 K. The solid and dashed lines indicate the alkyl chain carbon–(water)oxygen and –(water)hydrogen proximal correlation functions, respectively, evaluated from simulations of grafted alkyl chains in contact with water. The dots indicate the methane–(water)oxygen and –(water)hydrogen radial distribution functions, respectively, evaluated from simulations of a single methane molecule in water.

interfacial density profile, the proximal radial distribution function for carbon–(water)hydrogen differs from that for a solitary methane molecule in water (Ashbaugh and Paulaitis, 2001). The carbon–(water)hydrogen radial distribution function shows a primary peak at 3.5 Å with a weak shoulder at larger distances, which corresponds to water configurations in which the water hydrogen-bonding vector either straddles or points away from methane. The primary peak and shoulder in the proximal radial distribution function for carbon–(water)hydrogen, however, have merged into a single peak, shifted out to greater separations. These differences have been interpreted previously as changes in the three-dimensional water hydrogen-bonding network as a result of the inability of water to maintain the full network near a planar interface (Ashbaugh and Paulaitis, 2001).

The relationship between $g_{\text{prox}}(r)$ for water oxygen atoms (Fig. 1.11) and the oxygen atom interfacial density profile (Fig. 1.9) can be established by superposing these proximal radial distribution functions to model the conditional densities as

$$\rho_{\text{W}}(\boldsymbol{r}|\{\boldsymbol{r}_i\}) \approx \rho_{\text{w}} g_{\text{prox}}(\min_i\{|\boldsymbol{r} - \boldsymbol{r}_i|\}). \qquad (1.15)$$

These model conditional densities can then be averaged over the alkyl configurations sampled from simulation. The results of Fig. 1.9 show this exercise to be surprisingly successful; the proximal radial distribution function of Fig. 1.11 is sufficient to reconstruct accurately the density profile of Fig. 1.9.

We conclude that the proximal radial distribution function (Fig. 1.11) provides an effective deblurring of this interfacial profile (Fig. 1.9), and the deblurred structure is similar to that structure known from small molecule hydration results. The subtle differences of the $g_{prox}(r)$ for carbon–(water)hydrogen exhibited in Fig. 1.11 suggest how the thermodynamic properties of this interface, fully addressed, can differ from those obtained by simple analogy from a small molecular solute like methane; such distinctions should be kept in mind together to form a correct physical understanding of these systems.

Aside from conclusions specific to this physical system, we note that analysis of conditional properties achieves a strikingly simplified view of this statistical problem.

2

Statistical thermodynamic necessities

Our foremost goal here is to establish enough notation and a few pivotal relations that the following portions of the book can be understood straightforwardly. The following sections identify some basic thermodynamics and statistical thermodynamics concepts that will be used later. Many textbooks on thermodynamics and statistical mechanics are available to treat the basic results of this chapter in more detail; students particularly interested in solutions might consult Rowlinson and Swinton (1982).

Since this chapter is mostly notational, you might skip this chapter, but check back when you encounter notation that isn't immediately recognized. A glossary is provided; see p. xi.

2.1 The free energy and chemical potentials

The major goal of the theoretical developments will be a clear and practical access, on the basis of molecular information, to the chemical potential μ_α for a species of type α. The combination

$$\sum_\alpha \mu_\alpha n_\alpha = \mathcal{G}(n_1, n_2, \ldots, p, T) \tag{2.1}$$

is the Gibbs free energy, $\mathcal{G} = \mathcal{H} - T\mathcal{S}$, with $\mathcal{H}$ the enthalpy, T the absolute temperature, and $\mathcal{S}$ the entropy of the system. The property of extensivity, or first-order homogeneity discussed in standard textbooks (Callen, 1985), leads to the Gibbs–Duhem relation:

$$\sum_\alpha n_\alpha d\mu_\alpha = \mathcal{V}dp - \mathcal{S}dT. \tag{2.2}$$

The differential of the Gibbs free energy is

$$d\mathcal{G} = \mathcal{V}dp - \mathcal{S}dT + \sum_\alpha \mu_\alpha dn_\alpha. \tag{2.3}$$

Ideal results: chemical and conformational equilibrium

The simplest examples of chemical potentials occur when interactions between molecules are negligible. The chemical potentials are then evaluated as

$$\beta\mu_\alpha = \beta\mu_\alpha^{\text{ideal}} = \ln\left[\frac{\rho_\alpha}{\mathcal{Q}(n_\alpha = 1)/\mathcal{V}}\right]$$

$$\equiv \ln\left[\frac{\rho_\alpha \Lambda_\alpha^{\ 3}}{q_\alpha^{\text{int}}}\right]. \tag{2.4}$$

These relations will be established specifically in Chapter 3. $\beta^{-1} = kT$, where k is the Boltzmann constant, and Λ_α is the thermal de Broglie wavelength. $\mathcal{Q}(n_\alpha = 1) \equiv \mathcal{V}q_\alpha^{\text{int}}/\Lambda_\alpha^{\ 3}$ is the canonical ensemble partition function of a system comprising exactly one molecule of type α in a volume $\mathcal{V}$ at temperature T. The combination $\mathcal{V}/\Lambda_\alpha^{\ 3}$ is the contribution of translational motion to the partition function for the case of spatial homogeneity. q_α^{int}, dependent only on T, accounts for any degrees of freedom internal for a molecule of type α. Further details for a common example are discussed on p. 29. For the case that the system does, in fact, behave ideally we might alternatively write

$$\beta\mu_\alpha = \ln\left[\frac{x_j \beta p \Lambda_\alpha^3}{q_\alpha^{\text{int}}}\right], \tag{2.5}$$

with x_j the mole fraction of species j for this ideal case, $x_j p$ the partial pressure of species j, and p the total pressure. This form emphasizes the dependence of the chemical potentials on temperature, pressure, and composition.

On this basis we then consider conformational and chemical equilibrium in turn (Widom, 2002, see Chapter 3):

conformational. Consider A $\rightleftharpoons$ A′, and the equilibrium ratio for the ideal case. Then $\mu_A = \mu_{A'}$ leads to

$$K^{(0)}(T) = \frac{\rho_{A'}}{\rho_A} = \frac{q_{A'}^{\text{int}}}{q_A^{\text{int}}}, \tag{2.6}$$

where q_A^{int} and $q_{A'}^{\text{int}}$ are the single-molecule internal partition functions for a molecule in conformations A and A′, respectively.

chemical. Consider a chemical transformation, such as

$$n_A A + n_B B \rightleftharpoons n_C C + n_D D. \tag{2.7}$$

Here $n_C \mu_C + n_D \mu_D = n_A \mu_A + n_B \mu_B$ leads to

$$K^{(0)}(T) \equiv \frac{\rho_C{}^{n_C} \rho_D{}^{n_D}}{\rho_A{}^{n_A} \rho_B{}^{n_B}}$$

$$= \frac{\left(q_C^{\text{int}}/\Lambda_C{}^3\right)^{n_C} \left(q_D^{\text{int}}/\Lambda_D{}^3\right)^{n_D}}{\left(q_A^{\text{int}}/\Lambda_A{}^3\right)^{n_A} \left(q_B^{\text{int}}/\Lambda_B{}^3\right)^{n_B}}. \tag{2.8}$$

We emphasize that these results are for the ideal case and we will seek a natural generalization later.

Mixing free energies

When considering mixtures of two or more components, the thermodynamics of the mixtures is often cast in the form of changes in quantities on mixing. The change in Gibbs free energy on mixing at constant temperature and pressure is defined as

$$\Delta \mathcal{G}_{\text{mix}} = \mathcal{G} - \sum_\alpha n_\alpha \bar{\mu}_\alpha, \tag{2.9}$$

with $\bar{\mu}_\alpha$ the chemical potential of pure component α. Using Eqs. (2.1) and (2.5) this free energy change on mixing for the ideal case is found to be

$$\Delta \mathcal{G}_{\text{mix}}^{\text{ideal}} = kT \sum_\alpha n_\alpha \ln x_\alpha. \tag{2.10}$$

An important consequence is

$$\Delta \mathcal{S}_{\text{mix}}^{\text{ideal}} = -k \sum_\alpha n_\alpha \ln x_\alpha, \tag{2.11}$$

which is obtained from

$$\left(\frac{\partial \Delta \mathcal{G}_{\text{mix}}}{\partial T}\right)_{p,\boldsymbol{n}} = -\Delta \mathcal{S}_{\text{mix}}. \tag{2.12}$$

For an ideal solution, the enthalpy and volume changes of mixing are zero.

Partial molar quantities

The chemical potential is the partial molar Gibbs free energy. Partial molar quantities figure importantly in the theory of solutions and are defined at constant temperature and pressure; thus, the Gibbs free energy is a natural state function for their derivation. As an example, the partial molar volume is found from the Maxwell relation

$$\left(\frac{\partial \mu_\alpha}{\partial p}\right)_{T,\boldsymbol{n}} = \left(\frac{\partial \mathcal{V}}{\partial n_\alpha}\right)_{T,n_{\gamma \neq \alpha}} = v_\alpha(T, p, \boldsymbol{n}). \tag{2.13}$$

Then, the total volume is

$$\mathcal{V} = \sum_\alpha n_\alpha v_\alpha, \tag{2.14}$$

which is analogous to Eq. (2.1), p. 23. This is another illustration of the first-order homogeneity property, of $\mathcal{V}$ in this case.

Statistical thermodynamics

A foundational quantity is the canonical partition function

$$e^{-\beta \mathcal{A}} = Tr\left[e^{-\beta \mathbb{H}}\right]. \tag{2.15}$$

$\mathbb{H} = \mathbb{K} + \mathbb{V}$ is the Hamiltonian, the sum of kinetic ($\mathbb{K}$) and potential ($\mathbb{V}$) energies – operators in the case of quantum mechanics. $\mathcal{A}(\boldsymbol{n}, \mathcal{V}, T) = \mathcal{E} - T\mathcal{S}$ is the Helmholtz free energy, and $Tr[\mathbb{B}] = \sum_j \mathbb{B}_{jj}$ indicates a *trace* operation as explained in the standard textbooks, e.g. Münster (1969, Section 2.16) or Feynman (1972, Section 2.3).

One view of this trace operation is that the usual phase space integral may be obtained by representing the thermal density matrix $e^{-\beta \mathbb{H}}$ in plane-wave momentum states, and performing the trace in that state space (Landau *et al.*, 1980, Section 33. "Expansion in powers of $\hbar$"). Particle distinguishability restrictions are essential physical requirements for that calculation. In this book we will confine ourselves to the Boltzmann–Gibbs case so that $e^{-\beta \mathcal{A}} = \mathcal{Q}(\boldsymbol{n}, \mathcal{V}, T)/\boldsymbol{n}!$, since the

$$\boldsymbol{n}! \equiv \prod_\alpha n_\alpha! \tag{2.16}$$

that results from Boltzmann statistics is a crucial feature of subsequent maneuvers. We will typically omit the explicit display of the $\mathcal{V}, T$ dependence $\mathcal{Q}(\boldsymbol{n}) = \mathcal{Q}(\boldsymbol{n}, \mathcal{V}, T)$.

We will use the *absolute activities* $z_\alpha \equiv e^{\beta \mu_\alpha}$. Then the grand canonical partition function will be expressed as

$$\sum_{n \geq 0} \mathcal{Q}(\boldsymbol{n}) \frac{\boldsymbol{z}^n}{\boldsymbol{n}!} = e^{\beta p \mathcal{V}}, \tag{2.17}$$

a function of $(\boldsymbol{z}, \mathcal{V}, T)$ with $\boldsymbol{z} = \{z_1, z_2, \ldots\}$ the set of absolute activities; this uses the notation

$$\boldsymbol{z}^n \equiv \prod_\alpha z_\alpha^{n_\alpha}. \tag{2.18}$$

The sum in Eq. (2.17) is over all terms with integer $n_\alpha \geq 0$.

Another element in this whirlwind notational tour of statistical mechanics is a more explicit notation for averages. A canonical ensemble average of a phase function O_n will be denoted by

$$\langle O_n \rangle_{\mathrm{C}} = e^{\beta A} Tr \left(e^{-\beta \mathbb{H}} O_n \right). \tag{2.19}$$

Then a *grand* canonical ensemble average of the same property is

$$\langle O \rangle_{\mathrm{GC}} = e^{-\beta p \mathcal{V}} \sum_{n \geq 0} \langle O_n \rangle_{\mathrm{C}} \, \mathcal{Q}(n) \frac{z^n}{n!}. \tag{2.20}$$

Here $e^{\beta p \mathcal{V}}$ is the grand canonical partition function and serves as a normalizing factor in Eq. (2.20). Many useful statistical results don't depend on the ensemble used in evaluating an average, typically including averages of quantities not constrained by specification of the ensemble. In those cases, we will typically not use the subscripts that explicitly indicate the ensemble used.

Taking the derivatives of this grand canonical partition function with respect to the z_α with T and $\mathcal{V}$ fixed produces

$$\beta \mathrm{d}p = \sum_\alpha \langle \rho_\alpha \rangle \, \mathrm{d} \ln z_\alpha, \tag{2.21}$$

the *statistical* thermodynamic result that corresponds to Eq. (2.2) with $\mathrm{d}T = 0$.

Fluctuations

An important element in the discussions that follow, and in physical understanding, is the classic connection of fluctuations and susceptibilities with thermodynamic second derivatives. First derivatives of these thermodynamic potentials yield the composition of a system

$$z \left(\frac{\partial \beta p \mathcal{V}}{\partial z} \right)_{\beta, \mathcal{V}} = \langle n \rangle_{\mathrm{GC}}. \tag{2.22}$$

A second derivative produces

$$z \left(\frac{\partial \langle n \rangle_{\mathrm{GC}}}{\partial z} \right)_{\beta, \mathcal{V}} = \langle n^2 \rangle_{\mathrm{GC}} - \langle n \rangle_{\mathrm{GC}}^2 = \langle \delta n^2 \rangle_{\mathrm{GC}}. \tag{2.23}$$

Using $\rho = n/\mathcal{V}$, and remembering that $\mathrm{d} \ln z = \beta \mathrm{d}\mu = \beta \mathrm{d}p / \langle \rho \rangle_{\mathrm{GC}}$ for a one component system, this can be expressed as

$$\left(\frac{\partial \langle \rho \rangle_{\mathrm{GC}}}{\partial \beta p \mathcal{V}} \right)_{\beta, \mathcal{V}} = \frac{\langle \delta \rho^2 \rangle_{\mathrm{GC}}}{\langle \rho \rangle_{\mathrm{GC}}}. \tag{2.24}$$

A thermodynamic interpretation of this relation is that the left side is inversely proportional to the system volume $\mathcal{V}$. This relation then says that the fluctuations of the density of the material in a volume $\mathcal{V}$ are small for large $\mathcal{V}$. This can be true for large $\mathcal{V}$ if densities in different subvolumes become uncorrelated for sufficiently distant subvolumes.

Details of $\mathcal{Q}(n)$ for a typical case

Since $A = \sum_\alpha \mu_\alpha n_\alpha - p\mathcal{V}$, we can use Eq. (2.4), and supply the ideal gas equation of state to display $\mathcal{Q}(n)$ for that case of an ideal gas:

$$\beta A^{\text{ideal}} = \sum_\alpha n_\alpha \left(\ln \left[\frac{\rho_\alpha \Lambda_\alpha^3}{q_\alpha^{\text{int}}} \right] - 1 \right) \equiv - \ln \left[\mathcal{Q}(n)^{\text{ideal}} / n! \right]. \tag{2.25}$$

The physical point is that this result is additive over molecules, and therefore we conclude that $\mathcal{Q}(n)^{\text{ideal}}/n!$ is multiplicative over molecules.

Such a form can arise when $\mathbb{H}$ is additive over molecules. Indeed, this ideal gas result is obtained when the interactions between molecules, described in the potential energy $\mathbb{V}$, are negligible. This will be valid if the density of the system is sufficiently low; then the classical-limit Boltzmann–Gibbs $n!$ will also be satisfactory.

We now consider a case more typical than that of an ideal gas, though still special. We write

$$\frac{\mathcal{Q}(n)}{n!} = \frac{\mathcal{Q}(n)^{\text{ideal}}}{n!} \left(\frac{\mathcal{Q}(n)}{\mathcal{Q}(n)^{\text{ideal}}} \right), \tag{2.26}$$

and assume that a classically structured model

$$\frac{\mathcal{Q}(n)}{\mathcal{Q}(n)^{\text{ideal}}} = \left\langle e^{-\beta U(\mathcal{N})} \right\rangle_{\text{ideal}} \tag{2.27}$$

is satisfactory. Here $U(\mathcal{N})$ is the classical potential energy function describing intermolecular interactions at configuration $\mathcal{N}$, and the brackets $\langle \ldots \rangle_{\text{ideal}}$ indicate an average over configurations $\mathcal{N}$ sampled for the ideal gas case. This result assumes, as is the case for classical mechanics, that the kinetic energy ($\mathbb{K}$) and intramolecular potential energy contributions are separable from the intermolecular potential energy contributions; those kinetic energy and intramolecular contributions do constitute $\mathcal{Q}(n)^{\text{ideal}}$.

Equation (2.27) is proposed as a formal specification of a partition function, useful for a typical case encountered in the molecular theory of solutions. It is not suggested to be a practical method of calculation. We have here obviously taken pains not to commit to any specific molecular coordinates, but such a typical formulation as Eq. (2.27) will be helpful in our subsequent formal development.

A specific consequence of such details, and a result that will be used repeatedly, follows from consideration of the change in a partition function due to a change in the interactions treated. From Eq. (2.27) we see that

$$-\delta \ln \mathcal{Q}(\boldsymbol{n}) = \frac{\left\langle \delta\left(\beta U\left(\mathcal{N}\right)\right) e^{-\beta U(\mathcal{N})} \right\rangle_{\text{ideal}}}{\left\langle e^{-\beta U(\mathcal{N})} \right\rangle_{\text{ideal}}} = \beta \left\langle \delta U\left(\mathcal{N}\right) \right\rangle_U. \tag{2.28}$$

The average indicated by $\langle \ldots \rangle_U$ is the thermal average with all the interactions implied by $U(\mathcal{N})$ fully involved.

Details of q_α^{int} and $s_\alpha^{(0)}(\mathcal{R}^n)$ for a typical case

The internal partition function of a molecule, q_α^{int}, and the conformational distribution function, $s_\alpha^{(0)}(\mathcal{R}^n)$, discussed above on p. 24 and p. 18, respectively, were left unspecified initially. This is because different choices would be made for treating these objects in different physical situations. Nevertheless, it is helpful to give further definite details for a typical case, and to indicate some of the further possibilities.

The typical case we consider is the most natural initial treatment of molecules of non-trivial spatial extent, and with some conformational flexibility treated by molecular-mechanics force fields. We will here denote that mechanical potential energy surface by $U_\alpha(\mathcal{R}^n)$; the molecular type is denoted by the subscript α here. The present description will utilize cartesian coordinates $\mathcal{R}^n = \{r_1, r_2, \ldots, r_n\}$ where the r_j locates the jth atom taken as distinguishable but with definite, assigned atom-type. We assume that $U_\alpha(\mathcal{R}^n)$ is unchanged when atoms of the same type exchange locations.

One preliminary issue is to establish the configurational domain corresponding to recognition of a molecule of type α (Lewis *et al.*, 1961, see discussion, pp. 307–8). To accomplish this we use an indicator function $b_\alpha(\mathcal{R}^n)$ which is one (1) for configurations $\mathcal{R}^n$ recognized as forming a molecule of type α, and is zero (0) otherwise. The specification of $b_\alpha(\mathcal{R}^n)$, typically guided by physical considerations, involves some arbitrariness. But the definiteness of $b_\alpha(\mathcal{R}^n)$ means that physical predictions of a theory or computation can be examined for sensitivity to physically reasonable but arbitrary choices.

With this setup, a natural, semi-classical (Münster, 1969) specification is

$$\frac{\mathcal{V} q_\alpha^{\text{int}}}{\Lambda_\alpha^3} = \frac{1}{n!} \left(\prod_{j=1}^{j=n} \int_{\mathcal{V}} \frac{\mathrm{d}r_j}{\Lambda_j^3} \right) e^{-\beta U_\alpha(\mathcal{R}^n)} b_\alpha(\mathcal{R}^n). \tag{2.29}$$

Here $n!$ is the order of the group of permutations of atoms of the same atom type; this is the usual classical-limit treatment of the physical fact of indistinguishability of atoms of the same type. The Λ_j are defined by

$$\frac{1}{\Lambda_j} = \int_{-\infty}^{\infty} \exp\left[-\beta \frac{p^2}{2m_j}\right] \frac{dp}{h}, \tag{2.30}$$

where m_j is the mass of atom j. Λ_α has the same definition but with $m_\alpha = \sum_{j=1}^{j=n} m_j$, the total mass of a molecule of type α. The same physical description of $s_\alpha^{(0)}(\mathcal{R}^n)$ would be

$$\frac{s_\alpha^{(0)}(\mathcal{R}^n)}{V} = \frac{e^{-\beta U_\alpha(\mathcal{R}^n)} b_\alpha(\mathcal{R}^n)}{\left(\prod_{j=1}^{j=n} \int_V dr_j\right) e^{-\beta U_\alpha(\mathcal{R}^n)} b_\alpha(\mathcal{R}^n)}. \tag{2.31}$$

In terms of q_α^{int} this is

$$s_\alpha^{(0)}(\mathcal{R}^n) = \frac{e^{-\beta U_\alpha(\mathcal{R}^n)} b_\alpha(\mathcal{R}^n)}{n! \left(\frac{q_\alpha^{\text{int}}}{\Lambda_\alpha^3}\right) \prod_{j=1}^{j=n} \Lambda_j^3}. \tag{2.32}$$

The spatial integrations of Eq. (2.29), involving the coordinates of each atom, cover the volume V of the container. A reasonable definition of $b_\alpha(\mathcal{R}^n)$ would require bound atoms to be within a molecular length of each other, and should not depend on the location of a center, such as the center of mass, of the molecule within the volume. Then q_α^{int} will be independent of V asymptotically for large V. Similarly $s_\alpha^{(0)}(\mathcal{R}^n)$ will be independent of V for large V. The factors of Λ_α^3, in Eq. (2.29) is purely for notational convenience, and disappears in formulae such as Eq. (3.4), p. 33. $q_\alpha^{\text{int}} = 1$ for an atomic case.

This bare-bones treatment can be elaborated in considerable variety. Coordinate changes are the most primitive possibility; dynamical constraints that are popular in molecular dynamics simulation packages can be addressed from this point of view. Slightly trickier are cases for which $U_\alpha(\mathcal{R}^n)$ treats some atoms of the same type differently for numerical convenience; in that case the $n!$ should be thought through again, and might appear differently in a final formula. Though this expression is a limiting classical formulation, quantum statistical mechanical treatments based upon path integration (Feynman and Hibbs, 1965) can be pursued from this starting point. The traditional separate treatment of vibrations and rotations for small molecules is a hybrid case of making coordinate changes, then treating some coordinates essentially classically, but others in a convenient quantum mechanical approximation. This discussion does not exhaust the possibilities.

Exercises

2.1 Derive an expression for the partial molar entropy analogous to Eq. (2.13).

2.2 Use the ideal result Eq. (2.4) to derive the ideal gas equation of state on the basis of the Gibbs–Duhem Eq. (2.21).

2.3 Work out the statistical thermodynamic derivation of the Gibbs–Duhem Eq. (2.2), as for Eq. (2.21) but with $dT \neq 0$, and discuss the identification of the coefficient of dT.

3

Potential distribution theorem

The quantity of primary interest in our thermodynamic construction is the partial molar Gibbs free energy or chemical potential of the solute in solution. This chemical potential depends on the solution conditions: the temperature, pressure, and solution composition. A standard thermodynamic analysis of equilibrium concludes that the chemical potential in a local region of a system is independent of spatial position. The ideal and excess contributions to the chemical potential determine the driving forces for chemical equilibrium, solute partitioning, and conformational equilibrium. This section introduces results that will be the object of the following portions of the chapter, and gives an initial discussion of those expected results.

For a simple solute with no internal structure, i.e. no intramolecular degrees of freedom and therefore $q_\alpha^{\text{int}} = 1$, this chemical potential can be expressed as

$$\beta \mu_\alpha = \ln \rho_\alpha \Lambda_\alpha^3 + \beta \mu_\alpha^{\text{ex}}. \tag{3.1}$$

Since the density ρ_α appears in a dimensionless combination here, the concentration dependence of the chemical potential comes with a choice of concentration units. The first term on the right side of Eq. (3.1) expresses the colligative property of dilute solutions that the thermodynamic activity of the solute, $z_\alpha \equiv e^{\beta \mu_\alpha}$, is proportional to its concentration, ρ_α. The excess chemical potential accounts for intermolecular interactions between the solution molecules, and is given by the potential distribution theorem (Widom, 1963; 1982):

$$\beta \mu_\alpha^{\text{ex}} = -\ln \left\langle e^{-\beta \Delta U_\alpha} \right\rangle_0, \tag{3.2}$$

where ΔU_α is the potential energy of the interactions between the solution and a distinguished solute molecule of type α, and the brackets indicate thermal averaging over all solution configurations of the enclosed Boltzmann factor. The subscript zero emphasizes that this average is performed in the absence of interactions between the distinguished solute molecule and the solution. Note that the solution can be a complex mixture of species. For a dilute protein solution, the solution would typically include buffer, added salt, counter-ions, water, and perhaps cosolvents.

For a solute with internal degrees of freedom, the chemical potential is given by (Pratt, 1998)

$$\beta\mu_\alpha = \ln\left[\frac{\rho_\alpha}{\mathcal{Q}(n_\alpha = 1)/\mathcal{V}}\right] - \ln\left\langle\!\left\langle e^{-\beta\Delta U_\alpha}\right\rangle\!\right\rangle_0, \tag{3.3}$$

where $\mathcal{Q}(n_\alpha = 1) \equiv \mathcal{V}q_\alpha^{\text{int}}/\Lambda_\alpha{}^3$ is the partition function for a single solute molecule – see Eq. (2.4), p. 24, and discussion on p. 29. The double brackets indicate averaging over the thermal motion of the solute and the solvent molecules under the condition of no solute–solvent interactions – see Fig. 1.8, p. 17.

The probability that a solute molecule will adopt a specific conformation in solution is related to its chemical potential. This probability density function could be addressed in terms of the number density of solute molecules in conformation $\mathcal{R}^n$, $\rho_\alpha(\mathcal{R}^n)$ – see Fig. 1.8, p. 17:

$$\rho_\alpha(\mathcal{R}^n) = s_\alpha^{(0)}(\mathcal{R}^n)\left(\frac{z_\alpha q_\alpha^{\text{int}}}{\Lambda_\alpha{}^3}\right)\left\langle e^{-\beta\Delta U_\alpha} | \mathcal{R}^n\right\rangle_0, \tag{3.4}$$

where $s_\alpha^{(0)}(\mathcal{R}^n)$ is the normalized probability density for solute conformation $\mathcal{R}^n$ in the absence of interactions with the solvent, and the brackets indicate the thermodynamic average over solvent configurations with the solute fixed in conformation $\mathcal{R}^n$. Equation (3.3) is obtained from Eq. (3.4) by averaging over solute conformations, recognizing that $\rho_\alpha(\mathcal{R}^n)$ is normalized by the total number of solute molecules, and remembering the adopted normalization discussed with Eq. (1.13), p. 18.

Partition function perspective

Equations (3.2) and (3.3) relate intermolecular interactions to measurable solution thermodynamic properties. The excess chemical potential is obtained from

$$e^{-\beta\mu_\alpha^{\text{ex}}} = \left\langle\!\left\langle e^{-\beta\Delta U_\alpha}\right\rangle\!\right\rangle_0 = \int \mathcal{P}_\alpha^{(0)}(\varepsilon)e^{-\beta\varepsilon}\,d\varepsilon, \tag{3.5}$$

which introduces the probability density

$$\mathcal{P}_\alpha^{(0)}(\varepsilon) = \left\langle\!\left\langle \delta(\varepsilon - \Delta U_\alpha)\right\rangle\!\right\rangle_0. \tag{3.6}$$

The Boltzmann factor, $e^{-\beta\Delta U_\alpha}$, preferentially weights the low-energy tail of the distribution. This amounts to reweighting $\mathcal{P}_\alpha^{(0)}(\varepsilon)$, giving higher probabilities to those solvent configurations that are more favorable for solute placement. We will later develop the point that $e^{-\beta(\varepsilon-\mu_\alpha^{ex})}\mathcal{P}_\alpha^{(0)}(\varepsilon)$ is the properly normalized distribution of ε with all interactions fully assessed; this relation reduces Eq. (3.5) to a normalization condition. Equation (3.5) suggests the evaluation of, for example, a canonical partition function

$$\int \mathcal{P}_\alpha^{(0)}(\varepsilon)e^{-\beta\varepsilon}d\varepsilon \leftrightarrow \sum_E \Omega(E)e^{-\beta E}. \tag{3.7}$$

The summation is over all energy levels and $\Omega(E)$ is the number of thermodynamic states with energy E. See Fig. 3.1.

The chemical potentials sought are intensive properties of the system, in the usual thermodynamic language (Callen, 1985). Furthermore, the magnitude of ΔU_α is of molecular order, and the calculation of ΔU_α will depend on information about solution conditions in the neighborhood of the distinguished molecule. Thus, we expect the probability distribution functions of Eq. (3.6) to be independent of system size for thermodynamically large systems. This facilitates development of physical models for these distribution functions. For a case of long-ranged interactions, the neighborhood would be more extended than for a case of short-

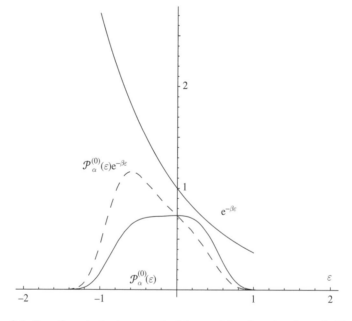

Figure 3.1 Functions in the integrand of the partition function formula Eq. (3.7). The lower solid curve labeled $\mathcal{P}_\alpha^{(0)}(\varepsilon)$ is the probability distribution of solute–solvent interaction energies sampled from the uncoupled ensemble of solvent configurations. The dashed curve is the product of this distribution and the exponential Boltzmann factor, $e^{-\beta\varepsilon}$, is the upper solid curve. See Eqs. (3.5) and (3.7).

ranged interactions. But the point remains that an accurate determination of this partition function should be possible based upon the molecular details of the solution in the vicinity of the solute.

The usual view of the PDT Eq. (3.3) is that it is a *particle insertion formula* (Valleau and Torrie, 1977; Frenkel and Smit, 2002). This is a view for implementing the indicated averaging: solution configurations are sampled without the distinguished molecule, and then the distinguished molecule is imposed upon the solution. The natural statistical estimate of the Boltzmann factor then provides the approximate evaluation of this partition function. This procedure has been extensively used, and naturally it will fail as a practical matter if the variance of the estimator is too large.

We've written Eq. (3.3) generally to suggest an alternative view. The solution could be imposed upon the distinguished molecule rather than the other way round. This may seem a trivial distinction, but it suggests a theoretical procedure in which the solution is built up around the distinguished molecule. This view is at the heart of quasi-chemical theories taken up later, and is analogous to the view of many-electron atoms based upon an *Aufbau* principle (Pauli, 1925; Karplus and Porter, 1970). The examples that we encounter argue that this build-up procedure has broader utility than the particle insertion idea. General derivations of the PDT needn't involve particle insertion concepts, and those general derivations are probably the most useful ones (Widom, 1978; 1982; Paulaitis and Pratt, 2002).

Exercises

3.1 The Ostwald partition coefficient, L, is a widely used and physically intuitive measure of gas solubilities and oil–water partition coefficients. It is defined as the ratio of concentrations of a solute between two phases at equilibrium. These two phases can be the ideal gas and a liquid phase, in which case the Ostwald partition coefficient gives the gas solubility, or two immiscible liquids – e.g., oil and water – in which case L is an oil–water partition coefficient. For the gas solubility of component 2 in liquid 1,

$$L \equiv \frac{\rho_2^{\text{liq}}}{\rho_2^{\text{vap}}}. \tag{3.8}$$

Derive a general expression for the Ostwald partition coefficient in terms of the excess chemical potential of the solute (component 2) in each of the two phases of interest. Table 3.1 gives free energies of hydration or excess chemical potentials for several n-alkanes in liquid water relative to the ideal gas at 25 °C. These free energies were computed from molecular simulations

Table 3.1 *Hydration free energies* $(kcal mol^{-1})$ *of n-alkanes and cavity analogs of the n-alkanes at 25°C calculated from molecular simulation*

solute	*n*-alkane	cavity analog
methane	2.63 (0.11)	6.65 (0.15)
ethane	2.57 (0.12)	9.07 (0.16)
propane	3.30 (0.13)	11.60 (0.17)
n-butane	3.51 (0.13)	13.84 (0.18)
n-pentane	3.69 (0.14)	15.99 (0.19)
n-hexane	3.92 (0.15)	18.24 (0.20)

(Ashbaugh *et al.*, 1999)

in which the alkanes were modeled using Lennard–Jones potential parameters for the -CH$_3$ and -CH$_2$-groups. Hydration free energies of cavity analogs of the *n*-alkanes at 25°C, computed in a second set of simulations, are also given in the table. The cavity analogs were designed to characterize just the excluded volume interactions of the alkanes with water. Using Eq. (3.3), devise a two-step process for *n*-alkane dissolution in water based on these two hydration free energies as input. The averaging formula derived in Section 3.3 below may prove helpful. Pay particular attention to the sign and the magnitude of the free energy change for each step, and explain how the free energy in each step would change if, instead, the alkane was transferred from another liquid *n*-alkane – e.g., *n*-octane – to water. Estimate the Ostwald partition coefficient for *n*-hexane in water relative to liquid *n*-octane at 25°C.

3.2 Aqueous solutions of poly(ethylene oxide), (-CH$_2$OCH$_2$-)$_m$, (PEO) exhibit unusual phase behavior in the sense that the polymer is soluble in water at room temperature, but phase separates at higher temperatures. Molecular interpretations of this unusual behavior implicate the effect of hydration on the conformational equilibria of the polymer chains based on studies of conformational equilibria of small-molecule analogs of PEO; e.g., 1,2 dimethoxyethane (DME), CH$_3$OCH$_2$CH$_2$OCH$_3$. Note that DME conformations are characterized by three consecutive dihedral angles centered on the O-C, C-C, and C-O bonds along the chain backbone, which can be found in the *trans* (t, ~ 180°), *gauche* (g, ~ +60°), and *gauche'* (g', ~ −60°) states. Table 3.2 gives hydration free energies and ideal gas intramolecular energies for the five most abundant chain conformations of DME. Show that *tgg'* is the most populated while *tgg* and *ttg* are the least populated conformations in the ideal gas based

Table 3.2 *Ab initio optimized geometries, dipole moments, and intramolecular energies, ΔE^*, of 1,2 dimethoxyethane[1] and free energies of hydration, μ_α^{ex}, at 25°C predicted from a continuum solvent model supported by molecular simulations[2]*

[1] Jaffe *et al.*, 1993
[2] Ashbaugh, 1998

conformation	conformational degeneracy[a]	dipole moment (Debye)	$\Delta E^*(\mathrm{kcal\,mol}^{-1})$	$\mu_\alpha^{\mathrm{ex}}(\mathrm{kcal\,mol}^{-1})$
ttt	1	0	0	0
tgt	2	1.52	0.14	−1.08
tgg	4	2.67	1.51	−1.71
tgg'	4	1.65	0.23	−0.26
ttg	4	1.93	1.43	−0.41

[a] The conformational degeneracy is the number of identical chain conformations corresponding to the given sequence of dihedral angles.

on energetics and the number of identical chain conformations corresponding to the given sequence of dihedral angles. Using *ttt* as the reference, compare the population densities of the other four conformations of DME in water relative to the ideal gas. What is the average dipole moment of DME in water compared with the ideal gas?

3.3 Consider the expression for the canonical partition function of a pure fluid:

$$Q(N, V, T) = \frac{1}{N!} \left(\frac{q_\alpha^{\mathrm{int}} V}{\Lambda_\alpha^3} \right)^N \langle e^{-\beta U(\mathcal{N})} \rangle_{\mathrm{ideal}}, \qquad (3.9)$$

in which $U(\mathcal{N})$ is the total potential energy of interactions among the N molecules in solution, see Eq. (2.27). Derive an expression for the chemical potential of this fluid using the standard thermodynamic relation,

$$\mu \equiv \left(\frac{\partial \mathcal{A}}{\partial N} \right)_{V, T}. \qquad (3.10)$$

In evaluating this derivative, consider adding a distinguished molecule to the N molecules in solution, $\mu \approx \frac{\Delta \mathcal{A}}{\Delta N}$. The total potential energy might then be expressed as

$$U(\mathcal{N} + 1) = U(\mathcal{N}) + \Delta U(\mathcal{N} + 1), \qquad (3.11)$$

where ΔU is the potential energy of interactions between the added molecule and the other N molecules in the fluid.

Table 3.3 *Tpρ data for liquid n-heptane along
the saturation curve from the normal boiling
point to the critical point.*

http://webbook.nist.gov/chemistry

temperature (K)	pressure (MPa)	$\rho(\text{g ml}^{-1})$
370.00	0.096840	0.61567
380.00	0.12910	0.60606
390.00	0.16918	0.59620
400.00	0.21825	0.58603
410.00	0.27758	0.57551
420.00	0.34850	0.56460
430.00	0.43240	0.55322
440.00	0.53074	0.54129
450.00	0.64507	0.52870
460.00	0.77704	0.51531
470.00	0.92841	0.50092
480.00	1.1011	0.48526
490.00	1.2972	0.46791
500.00	1.5191	0.44822
510.00	1.7697	0.42501
520.00	2.0524	0.39581
530.00	2.3710	0.35395
540.00	2.7277	0.26570

3.4 The van der Waals equation of state for a pure fluid is derived from the following canonical partition function:

$$Q(N, \mathcal{V}, T) = \frac{1}{N!} \left(\frac{q_\alpha^{\text{int}}}{\Lambda_\alpha^3} \right)^N (\mathcal{V} - Nb)^N \, e^{\beta N^2 a/\mathcal{V}}. \qquad (3.12)$$

In the context of van der Waals theory, a and b are positive parameters characterizing, respectively, the magnitude of the attractive and repulsive (excluded volume) intermolecular interactions. Use this partition function to derive an expression for μ_α^{ex}, the excess chemical potential of a distinguished molecule (the solute) in its pure fluid. Note that specific terms in this expression can be related to contributions from either the attractive or excluded-volume interactions. Use the *Tpρ* data given in Table 3.3 for liquid *n*-heptane along its saturation curve to evaluate the influence of these separate contributions on test-particle insertions of a single *n*-heptane molecule in liquid *n*-heptane as a function of density. In light of your results, comment on the statement made in the discussion above that the use of the potential distribution theorem to evaluate μ_α^{ex} depends on primarily local interactions between the solute and the solvent.

3.5 Once you have the chemical potential for the van der Waals model of the previous exercise, find the equation of state by integrating the Gibbs–Duhem relation. Compare your result with the equation of state obtained from the approximate partition function using

$$\left(\frac{\partial \mathcal{A}}{\partial \mathcal{V}}\right)_{T,N} = -p. \tag{3.13}$$

3.1 Derivation of the potential distribution theorem

Consider a macroscopic solution consisting of the solute molecules of interest and other species. Focus attention on a macroscopic subsystem of this solution. We will describe this subsystem on the basis of the grand canonical ensemble of statistical thermodynamics, accordingly specified by a temperature, volume, and the chemical potentials or, equivalently, absolute activities for all solution species. The species of interest will be identified with a subscript index α, so the average number of solute molecules in this subsystem is

$$\langle n_\alpha \rangle = e^{-\beta p \mathcal{V}} \sum_{n \geq 0} n_\alpha \mathcal{Q}(n) \frac{z^n}{n!}. \tag{3.14}$$

The average displayed in Eq. (3.14) is particularly relevant to our argument here. The summand factor n_α annuls terms with $n_\alpha = 0$ and permits the sum to start with $n_\alpha = 1$. The latter feature means that the overall result will involve an explicit leading factor of z_α. We are then motivated to examine the coefficient multiplying z_α. Of course, a determination of z_α establishes the thermodynamic property μ_α that we seek. To that end, we bring forward the explicit extra factor of z_α and write

$$\langle n_\alpha \rangle = z_\alpha e^{-\beta p \mathcal{V}} \mathcal{Q}(n_\alpha = 1) \sum_{n \geq 0} \left(\frac{\mathcal{Q}(n_1 \ldots [n_\alpha + 1] \ldots)}{\mathcal{Q}(n_\alpha = 1)}\right) \frac{z^n}{n!}. \tag{3.15}$$

Notice that we haven't written the n_α of Eq. (3.14) explicitly in the summand of Eq. (3.15). It has been absorbed in the $n!$, but its presence is reflected in the fact that the population is enhanced by one in the numerator that appears in the summand. "$n_1 \ldots [n_\alpha + 1] \ldots$" indicates the population *after* one molecule of species α is added to n. $\mathcal{Q}(n_\alpha = 1)$ is the canonical partition function with exactly $n_\alpha = 1$ and no other molecules; in our notation above

$$\mathcal{Q}(n_\alpha = 1) \equiv \frac{q_\alpha^{\text{int}} \mathcal{V}}{\Lambda_\alpha^3}. \tag{3.16}$$

We've positioned this factor out front in writing Eq. (3.15) so that the leading $n = 0$ term in the sum will be one (1) as suggested by the usual partition function sums. Notice also that the summand of Eq. (3.15) would adopt precisely the form of a grand canonical average, e.g., Eq. (2.20), p. 27, if we were to discover a factor of $\mathscr{Q}(n)$ for the population weight. Thus

$$\rho_\alpha = \frac{z_\alpha q_\alpha^{int}}{\Lambda_\alpha^{3}} \left\langle \frac{\mathscr{Q}(n_1 \ldots [n_\alpha+1]\ldots)}{\mathscr{Q}(n_\alpha=1)\,\mathscr{Q}(n)} \right\rangle_{GC}. \tag{3.17}$$

Finally, we observe that the numerator and denominator of the population averaged ratio of Eq. (3.17) are configurational integrals involving the same coordinates. The numerator integrand is the Boltzmann factor for the system with the single extra solute molecule; see, for example, Eq. (2.27), p. 28. In contrast, the denominator integrand is the product of the Boltzmann factors for that distinguished extra solute molecule and the rest of the solution, but uncoupled. Thus,

$$\frac{\rho_\alpha \Lambda_\alpha^{3}}{z_\alpha q_\alpha^{int}} = \left\langle\!\left\langle e^{-\beta \Delta U_\alpha} \right\rangle\!\right\rangle_0 \equiv e^{-\beta \mu_\alpha^{ex}}, \tag{3.18}$$

which combines both the population averaging with the canonical ensemble partition function. In emphasizing again that the interactions can be any physical ones, we note specifically that this derivation also covers the case with a physical external field.

3.2 Weak field limit

The case of a weak external potential energy field is generally important, and the simple result is also suggestive. We augment the potential energy to include an external field, $\varphi_\alpha(\mathscr{R}^n)$, and work out that result for the chemical potential, $\beta\mu_\alpha^{ex}$. The notations

$$\beta\mu_\alpha[\varphi] = \ln\left(\frac{\rho_\alpha \Lambda_\alpha^{3}}{q_\alpha^{int}[\varphi]}\right) - \ln\left\langle\!\left\langle e^{-\beta\Delta U_\alpha}\right\rangle\!\right\rangle_0[\varphi] \tag{3.19}$$

emphasize that the statistical quantities such as $\beta\mu_\alpha^{ex}[\varphi]$ depend on the function $\varphi_\alpha(\mathscr{R}^n)$. This is true also for the internal partition function $q_\alpha^{int}[\varphi]$, and evaluating that quantity will be our first step:

$$-\ln q_\alpha^{int}[\varphi] \approx -\ln q_\alpha^{int}[0] + \beta\langle\varphi\rangle_0. \tag{3.20}$$

This approximation uses the fundamental but elementary point that $\ln\langle e^\varepsilon\rangle \approx \langle\varepsilon\rangle$ if ε hardly varies. This specific calculation will arise again several times in our further study. Eq. (3.20) neglects nonlinear contributions caused by statistical variation in φ, and the average on the right is for the distinguished molecule without other interactions. The leading term is the result without the field present.

See Eq. (2.29), p. 29, for a typical detailed $q_\alpha^{int}[\varphi]$ for a similar calculation.

Now consider the contributions of Eq. (3.19) from intermolecular interactions. Comparing Eq. (3.17) and Eq. (3.18), this is seen to be the logarithm of a ratio of integrals. Simple proportionality factors cancel in forming the ratio. Then the denominator of that ratio is a partition function for the uncoupled $(N+1)$-molecule system, i.e., without interactions between the N-molecule solution and the distinguished molecule. The numerator is similarly proportional to the partition function for the physical $(N+1)$-molecule system. We thus write

$$-\ln\left\langle\!\!\left\langle e^{-\beta\Delta U_\alpha}\right\rangle\!\!\right\rangle_0[\varphi] \approx -\ln\left\langle\!\!\left\langle e^{-\beta\Delta U_\alpha}\right\rangle\!\!\right\rangle_0[0]$$
$$+\beta\left\langle\left(\sum_{j=1}^{N+1}\varphi(j)\right)\right\rangle - \beta\left\langle\!\!\left\langle\left(\sum_{j=1}^{N+1}\varphi(j)\right)\right\rangle\!\!\right\rangle_0. \qquad (3.21)$$

Again, the doubled brackets with subscript zero in the last term imply no inter-actions between the N-molecule solution and the distinguished molecule. Thus, one term in that sum will precisely cancel the field contribution of Eq. (3.20). In composing Eq. (3.19), we then write

$$\mu_\alpha[\varphi] \approx \mu_\alpha[0] + \left\{\left\langle\left(\sum_{j=1}^{N+1}\varphi(j)\right)\right\rangle - \left\langle\sum_{j=1}^{N}\varphi(j)\right\rangle\right\}. \qquad (3.22)$$

Finally, we note that the last combination is the change in the physical field potential energy upon incrementing the molecular number by one, so

$$\mu_\alpha[\varphi] \approx \mu_\alpha[0] + \frac{\partial}{\partial\rho_\alpha}\left(\langle\varphi\rangle\rho_\alpha\right). \qquad (3.23)$$

In contrast with the averaging in Eq. (3.20), the averaging here is for the fully coupled system.

Exercise

3.6 For the case considered by Eq. (3.23), explain why the external field makes a contribution $\langle\varphi\rangle\rho_\alpha$ to the Helmholtz free energy per unit volume, $\mathcal{A}/\mathcal{V}$. On this basis, give a thermodynamic derivation of Eq. (3.23) (Landau & Lifshitz, vol. 8).

3.3 Potential distribution theorem view of averages

Note how averages associated with the system including the solute of interest look from the perspective of the potential distribution theorem. The averaging $\langle\!\langle\ldots\rangle\!\rangle_0$ does not involve the solute–solution interactions, of course. Those are averages over the thermal motion of the molecule and solution when the energetic coupling between those subsystems is eliminated. Consider then the average of a configurational function $F(\mathcal{N}+1)$ in the physical case when the interactions between the distinguished molecule of type α and the solution are fully involved. We obtain this average by supplying the Boltzmann factor for the interactions between the distinguished molecule and the solution, as well as the proper normalization:

$$\langle F \rangle = \frac{\langle\langle e^{-\beta \Delta U_\alpha} F \rangle\rangle_0}{\langle\langle e^{-\beta \Delta U_\alpha} \rangle\rangle_0}. \tag{3.24}$$

This establishes that

$$P(\mathcal{N}+1) = \left(\frac{s_\alpha^{(0)}(\mathcal{R}^n)}{V} \right) P_B^{(0)}(\mathcal{N}) e^{-\beta(\Delta U_\alpha - \mu_\alpha^{\mathrm{ex}})} \tag{3.25}$$

is the properly normalized canonical configurational probability for the fully coupled system. Here $P_B^{(0)}(\mathcal{N})$ is the canonical distribution for the bath uncoupled to the distinguished molecule, and $s_\alpha^{(0)}(\mathcal{R}^n)$ is the configurational distribution function discussed on p. 17, Fig. 1.8. Notice that the normalization for the distribution of the $(N+1)$-molecule system, here completed by the quantity $\beta \mu_\alpha^{\mathrm{ex}}$, depends on the *distribution* of $\mathcal{N}$ and $\mathcal{R}^n$, and isn't calculated for specific configurations, $\mathcal{N}$ and $\mathcal{R}^n$.

Exercises

3.7 Derive Eq. (3.24) using the expression for a grand canonical average:

$$\langle F \rangle = e^{-\beta p V} \sum_{n \geq 0} F \mathcal{Q}(n) \frac{z^n}{n!}. \tag{3.26}$$

3.8 Use Eqs. (3.18) and (3.24) with the choice $F = e^{\beta \Delta U_\alpha}$ to produce the important *inverse* formula

$$\left\langle e^{\beta \Delta U_\alpha} \right\rangle = e^{\beta \mu_\alpha^{\mathrm{ex}}}. \tag{3.27}$$

This and Eq. (3.18) provide different routes to calculating the solute excess chemical potential. Explain what those differences are.

3.9 Consider the result Eq. (3.25) and relate the distribution function of the binding energy for the distinguished solute in the actual, fully-coupled system $\mathcal{P}_\alpha(\varepsilon) = \langle \delta(\varepsilon - \Delta U_\alpha) \rangle$ to the distribution function $\mathcal{P}_\alpha^{(0)}(\varepsilon)$ of Eq. (3.6).

3.10 Consider a specific configuration $\mathcal{N}$ and determine the conditional distribution of the distinguished molecule $N+1$.

3.4 Ensemble dependence

Occasionally results alternative to the potential distribution theorem, expressed as Eq. (3.18), have been proposed (Frenkel, 1986; Shing and Chung, 1987; Smith, 1999). These alternatives typically consider statistical thermodynamic manipulations associated with a particular ensemble, and the distinguishing features of those alternative formulae are relics of the particular ensemble under consideration. On the other hand, the derivation above considered a macroscopic subsystem of a larger system avoiding specification of constraints on that larger system. The macroscopic subsystem is then appropriately analyzed on the basis of a grand canonical treatment of statistical thermodynamics, and the modifications specific to an ensemble aren't evident in the final formula. Here we discuss those distinctions and conclude that all these formulae should give the same result in the thermodynamic limit, and that the result Eq. (3.18) is to be preferred on the basis of its locality and relic-free appearance.

A specific example will be sufficient to fix these ideas. Consider calculations in an isothermal–isobaric ensemble (Shing and Chung, 1987; Smith, 1999):

$$e^{-\beta\mu_\alpha^{\text{ex}}} \leftarrow \frac{\langle\langle \mathcal{V}e^{-\beta\Delta U_\alpha}\rangle\rangle_0}{\langle\langle \mathcal{V}\rangle\rangle_0}, \tag{3.28}$$

for which the volume $\mathcal{V}$ fluctuates. In this formula $\mathcal{V}$ is a remnant of the ensemble considered. Let's analyze this distinction further. The derivation of Eq. (3.18) used the grand canonical ensemble to treat a system of definite volume, physically a subsystem of a larger system. To interpret the right side of Eq. (3.28) for a simulation in an ensemble in which the volume may fluctuate, we decide that the rule of averages Eq. (3.24), must still apply in the form

$$\frac{\langle\langle \mathcal{V}e^{-\beta\Delta^{\prime}U_\alpha}\rangle\rangle_0}{\langle \mathcal{V}\rangle_0} = \langle\langle e^{-\beta\Delta\mu_\alpha}\rangle\rangle_0 \langle \mathcal{V}\rangle$$

with $\langle \mathcal{V}\rangle$ the physical volume of the solution plus an additional molecule. With this guidance in interpreting the brackets $\langle\langle\cdots\rangle\rangle_0$, we write

$$\approx \langle\langle e^{-\beta\Delta\mu_\alpha}\rangle\rangle_0 \left(1 + \frac{1}{\langle \mathcal{V}\rangle_0}\left[\frac{\partial\langle \mathcal{V}\rangle}{\partial n_\alpha}\right]_{T,p,n_{\gamma\neq\alpha}}\right) \tag{3.29}$$

so that

$$\frac{\langle\langle \mathcal{V}e^{-\beta\Delta U_\alpha}\rangle\rangle_0}{\langle \mathcal{V}\rangle_0} - \langle\langle e^{-\beta\Delta\mu_\alpha}\rangle\rangle_0 \approx \frac{1}{\langle \mathcal{V}\rangle_0}\left[\frac{\partial\mu_\alpha}{\partial p}\right]_{T,n}. \tag{3.30}$$

The difference displayed by Eq. (3.30) is negligible in the macroscopic limit.

Simulation calculations on finite systems will entail some error associated with the submacroscopic size considered (Lebowitz and Percus, 1961a; b; 1963). For example, periodic boundary conditions will influence molecular correlations to some extent (Pratt and Haan, 1981a; b). Support of a claim of accuracy would typically involve some practical investigation of the thermodynamic limit. A claim of preference for calculations in one ensemble over another is typically made first on the basis of convenience rather than on the basis of accuracy defined in some absolute way. Thus, advantages of practical accuracy for ensemble-specialized alternatives to Eq. (3.18) are not proven typically, and they are not necessary fundamentally.

But notice that $\mathcal{V}$, in this example, introduces a fluctuating global variable into a formula otherwise involving the more local quantity $e^{-\beta \Delta U_\alpha}$; thus, introduction of $\mathcal{V}$ is a nuisance for subsequent molecular theory. Since such alternatives are not fundamentally required, we don't consider them further.

3.5 Inhomogeneous systems

Let's consider inhomogeneous molecular systems and discuss the atomic densities that are not now spatially constant. We might consider a crystal clamped in a diffraction apparatus and ask about the average positions of molecules. Or we might consider a liquid–vapor two-phase system in a bottle with gravity positioning the liquid on the bottom. We will also later use these results to investigate densities conditional on the location of specific other atoms within our system. The average density pattern we seek is

$$\rho_\alpha(r) = \left\langle \sum_{j \in \alpha} \delta(r_j - r) \right\rangle. \tag{3.31}$$

For convenience, we first discuss this problem for the case that an atomic description of the liquid is satisfactory. The sum is over all atoms of type α. This density is zero for the case that there are no atoms of type α ($n_\alpha = 0$), so for the cases making a nonzero contribution, we distinguish a specific one of those atoms of type α present, say the first with location r_1, and use Eq. (3.24). Thus we can write

$$\rho_\alpha(r) = \langle n_\alpha \rangle \langle \delta(r_1 - r) \rangle = \langle n_\alpha \rangle \frac{\langle\!\langle e^{-\beta \Delta U_\alpha} \delta(r_1 - r) \rangle\!\rangle_0}{\langle\!\langle e^{-\beta \Delta U_\alpha} \rangle\!\rangle_0}. \tag{3.32}$$

As far as the averaging on the right end of Eq. (3.32) is concerned, interactions between the solute and the rest of the system aren't present. The spatial averaging of the quantity $\delta(r_1 - r)$ is then straightforward: it locates the solute at r and evaluates the probability density

$$\langle\!\langle \delta(r_1 - r) \rangle\!\rangle_0 = \frac{e^{-\beta \varphi_\alpha(r)}}{\mathcal{V} q_\alpha^{\text{int}} [\varphi_\alpha]}, \tag{3.33}$$

where now the normalizing $q_\alpha^{\text{int}}[\varphi_\alpha]$ includes effects of the applied field, but the factor V supplied here insures that $q_\alpha^{\text{int}}[\varphi_\alpha = 0] = 1$. Introducing the *conditional expectation* (p. 18), we have

$$\rho_\alpha(r) = \frac{\rho_\alpha e^{-\beta\varphi_\alpha(r)}}{q_\alpha^{\text{int}}[\varphi_\alpha]} \left(\frac{\left\langle e^{-\beta\Delta U_\alpha}|r_1 = r\right\rangle_0}{\left\langle\!\left\langle e^{-\beta\Delta U_\alpha}\right\rangle\!\right\rangle_0} \right). \tag{3.34}$$

For the average in the numerator, the solute is now definitely located at the point r, and the notation here is intended to convey that restriction. The indicated conditional expectation denotes that the spatial averaging involves only the thermal motion of the solution. Finally the elimination of the denominator produces the notable form

$$\beta\mu_\alpha[\varphi] = \ln\rho_\alpha(r)\Lambda_\alpha^3 + \beta\varphi_\alpha(r) - \ln\left\langle e^{-\beta\Delta U_\alpha}|r\right\rangle_0, \tag{3.35}$$

which is analogous to Eqs. (3.18) and (3.19), p. 40. The notation of a conditional expectation indicates that a distinguished particle is located at r. A consequence is that $q_\alpha^{\text{int}}[\varphi_\alpha]$ with its dependence on the external field φ_α does not appear. In contrast, $q_\alpha^{\text{int}}[\varphi_\alpha]$ does appear in Eqs. (3.18) and (3.19) because the double brackets there specify that the distinguished solute should sample all available positions. That distinction brings in $q_\alpha^{\text{int}}[\varphi_\alpha]$, with any field dependence, as an additional normalizing factor.

If the system is uniform, then this relation reduces to

$$\beta\mu_\alpha = \ln\rho_\alpha\Lambda_\alpha^3 - \ln\left\langle e^{-\beta\Delta U_\alpha}|r\right\rangle_0. \tag{3.36}$$

This indicates that the conditional mean $\left\langle e^{-\beta\Delta U_\alpha}|r\right\rangle_0$ is independent of placement of the distinguished atom, as expected.

The same derivation can be directed towards the important previewed result Eq. (3.4), p. 33, the general case for a molecule with internal flexibility. The analogous development would involve

$$\frac{\rho_\alpha(\mathcal{R}^n)}{\langle n_\alpha\rangle} = \left\langle\!\left\langle \delta(\mathcal{R}_1^n - \mathcal{R}^n)\right\rangle\!\right\rangle_0 \left(\frac{\left\langle e^{-\beta\Delta U_\alpha}|\mathcal{R}_1^n = \mathcal{R}^n\right\rangle_0}{\left\langle\!\left\langle e^{-\beta\Delta U_\alpha}\right\rangle\!\right\rangle_0} \right). \tag{3.37}$$

In this case $\left\langle\!\left\langle \delta(\mathcal{R}_1^n - \mathcal{R}^n)\right\rangle\!\right\rangle_0 = s_\alpha^{(0)}(\mathcal{R}^n)/V$ where the definition of $s_\alpha^{(0)}(\mathcal{R}^n)$ was given earlier as the normalized probability density for solute of type α in conformation $\mathcal{R}^n$ in the absence of interactions with the solution. Using Eq. (3.3), p. 33, this is

$$\rho_\alpha(\mathcal{R}^n) = s_\alpha^{(0)}(\mathcal{R}^n)\left(\frac{z_\alpha q_\alpha^{\text{int}}}{\Lambda_\alpha^3} \right)\left\langle e^{-\beta\Delta U_\alpha}|\mathcal{R}^n\right\rangle_0, \tag{3.38}$$

and

$$\beta\mu_\alpha = \ln\left[\frac{\rho_\alpha(\mathcal{R}^n)\Lambda_\alpha^3}{s_\alpha^{(0)}(\mathcal{R}^n)q_\alpha^{\text{int}}} \right] - \ln\left\langle e^{-\beta\Delta U_\alpha}|\mathcal{R}^n\right\rangle_0. \tag{3.39}$$

The factor of $q_\alpha^{\rm int}$ here balances a normalizing factor in $s_\alpha^{(0)}(\mathcal{R}^n)$; as in the transition from Eq. (3.34) to Eq. (3.35), for a specific case of $s_\alpha^{(0)}(\mathcal{R}^n)$ it would be most natural to eliminate that unnecessary explicit appearance of $q_\alpha^{\rm int}$.

If molecular densities were determined on the basis of Eq. (3.38), atomic densities might be evaluated by contraction of those results. Equation (3.38) provides a derivation of the previously mentioned conditional density of Eq. (3.4). This point hints at a physical issue that we note. As we have emphasized, the potential distribution theorem doesn't require simplified models of the potential energy surface. A model that implies chemical formation of molecular structures can be a satisfactory description of such molecular systems. Then, an *atomic* formula such as Eq. (3.35) is fundamentally satisfactory. On the other hand, if it is clear that atoms combine to form molecules, then a molecular description with Eq. (3.38) may be more convenient. These issues will be relevant again in the discussion of quasi-chemical theories in Chapter 7 of this book. This issue comes up in just the same way in the next section.

Exercises

3.11 The internal partition function $q_\alpha^{\rm int}[\varphi_\alpha]$ does not appear in Eq. (3.35), though it does appear explicitly in Eqs. (3.18) and (3.19), p. 40. Start with Eq. (3.35), and show that this implies Eq. (3.19). Give a physical statement and interpretation of this distinction.

3.12 Using Eq. (3.35), derive an expression for the density of an ideal gas as a function of height, z, in a gravitational field. Verify that the same expression is obtained from Eq. (3.23), the chemical potential in the presence of a weak external field.

3.13 Consider the excess chemical potential obtained for a definite configuration, $\mu_\alpha^{\rm ex}(\mathcal{R}^n)$, and let $s_\alpha(\mathcal{R}^n)$ be the conformational distribution for the fully coupled case so that $\rho_\alpha(\mathcal{R}^n) = \rho_\alpha s_\alpha(\mathcal{R}^n)$. Show that (Imai and Hirata, 2003)

$$\beta\mu_\alpha^{\rm ex} = \langle \beta\mu_\alpha^{\rm ex}(\mathcal{R}^n) \rangle + \left\langle \ln\left[\frac{s_\alpha(\mathcal{R}^n)}{s_\alpha^{(0)}(\mathcal{R}^n)}\right]\right\rangle. \tag{3.40}$$

Although a similar relation – but pointwise – follows from Eq. (3.39), Eq. (3.40) is analogous with the standard relations establishing, e.g., the Helmholtz free energy in terms of a full canonical probability distribution, such as $P_B^{(0)}(\mathcal{N})$ of p. 42. For an example see Reiss (1972, see Eq. (9)), and for follow-on work (Schlijper and Kikuchi, 1990; Singer, 2004). Again here the conclusion is that Eq. (3.39) should be seen as a partition function formula.

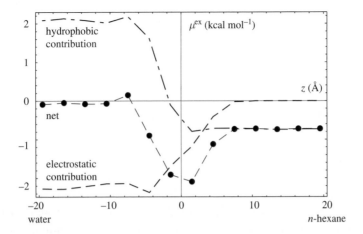

Figure 3.2 Variation of excess chemical potential of CH_3F as a function of distance of the carbon atom from the liquid water–hexane interface at 310 K (Pohorille and Wilson, 1996; Pratt and Pohorille, 2002). The hydrophobic contribution, obtained by eliminating electrostatic interactions, is the dot–dash curve and the electrostatic contribution is the dashed curve, lowest on the right. The water equimolar surface is at $z = 0$. The combination of these two contributions leads to interfacial activity for this simple solute.

3.14 Show how Eq. (3.39) reduces to Eq. (3.35) for the case of a mono-atomic "molecule." Hint: what is q_α^{int} in that case?

3.15 Figure 3.2 shows the excess chemical potential of CH_3F along the z-direction perpendicular to the water–hexane interface at 310 K obtained from molecular dynamics simulation using the particle insertion method. Water is on the left ($z < 0$) and hexane is on the right ($z > 0$). Calculate the Ostwald partition coefficient of CH_3F in the interfacial region relative to bulk water as a function of z. Give a physical interpretation of your result.

3.6 Reduced distribution functions

We next describe spatial distributions of pairs, triples, $\ldots$, m-tuples of atoms in our solution. The pair distribution is the usual quantity of interest, but we will find a general formula that will be more broadly useful.

In working towards a general formula we first make the observation that the derivation of Section 3.1 focuses on $\langle n_1 \rangle$ in a way that is simply generalized for the moment $\langle n_1(n_1 - 1) \rangle$. For example, direct analogy with Eq. (3.15) suggests

$$\langle n_1(n_1 - 1) \rangle = e^{-\beta pV} z_1^2 \mathcal{Q}(\{2, 0, \ldots\}) \sum_{n \geq 0} \left(\frac{\mathcal{Q}(n + \{2, 0, \ldots\})}{\mathcal{Q}(\{2, 0, \ldots\})} \right) \frac{z^n}{n!}. \quad (3.41)$$

The reason this is correct is that the combination $n_1(n_1 - 1)$ annuls the zeroth and first terms in n_1 that would arise. We emphasize this point by adopting the 'n-to-the-k-falling' notation

$$n^{\underline{k}} \equiv n(n-1)\ldots(n-k+1). \tag{3.42}$$

The results following here will consider distinguishing p-tuples of atoms from the bath, and we will need some further notation for the coupling energies that arise. We write

$$\Delta U^{(p)} \equiv U(\mathcal{N}+p) - U(\mathcal{N}) - U(1,\ldots,p) \tag{3.43}$$

for the change in energy when an N-molecule system and a p-tuple are coupled. None of these energies need be pair decomposable. The coupling energy in the formulae above would then be denoted here as $\Delta U = \Delta U^{(1)} = U(\mathcal{N}+1) - U(\mathcal{N}) - U(1)$. Using this notation, Eq. (3.41) can be rewritten as

$$\left\langle n_1^{\underline{2}} \right\rangle = z_1^{~2} \mathcal{Q}_2 \left\langle\!\!\left\langle e^{-\beta \Delta U^{(2)}} \right\rangle\!\!\right\rangle_0, \tag{3.44}$$

where the indicated average is over the thermal motion of the bath plus a distinguished pair of identical atoms with no coupling between these two subsystems, and $\mathcal{Q}_2/2$ is the canonical partition function for the pair, i.e. $\mathcal{Q}_2 = \mathcal{Q}(\{2,0,\ldots\})$ of Eq. (3.41). The more general relation is

$$\frac{\left\langle n_1^{\underline{p}} \right\rangle}{\left\langle n_1 \right\rangle^p} = \frac{\mathcal{Q}_p \left\langle\!\!\left\langle e^{-\beta \Delta U^{(p)}} \right\rangle\!\!\right\rangle_0}{\left(\mathcal{Q}_1 \left\langle\!\!\left\langle e^{-\beta \Delta U^{(1)}} \right\rangle\!\!\right\rangle_0\right)^p}. \tag{3.45}$$

We now consider spatial correlations between this distinguished pair of atoms by defining the pair correlation function, $g^{(2)}(\mathbf{r},\mathbf{r}')$, as follows:

$$\left(\frac{\langle n \rangle}{V}\right)^2 g^{(2)}(\mathbf{r},\mathbf{r}') = \left\langle n_{\underline{2}} \right\rangle \left\langle \delta(\mathbf{r}_1 - \mathbf{r})\delta(\mathbf{r}_2 - \mathbf{r}') \right\rangle, \tag{3.46}$$

or, in view of Eq. (3.45),

$$g^{(2)}(\mathbf{r},\mathbf{r}') = V^2 \frac{\mathcal{Q}_2 \left\langle\!\!\left\langle e^{-\beta \Delta U^{(2)}} \right\rangle\!\!\right\rangle_0}{\left(\mathcal{Q}_1 \left\langle\!\!\left\langle e^{-\beta \Delta U^{(1)}} \right\rangle\!\!\right\rangle_0\right)^2} \left\langle \delta(\mathbf{r}_1 - \mathbf{r})\delta(\mathbf{r}_2 - \mathbf{r}') \right\rangle. \tag{3.47}$$

The joint probability density for the positions of two specific atoms can be evaluated using the potential distribution theorem formula for averages, Eq. (3.24), p. 42:

$$\left\langle \delta(\mathbf{r}_1 - \mathbf{r})\delta(\mathbf{r}_2 - \mathbf{r}') \right\rangle = \frac{\left\langle\!\!\left\langle e^{-\beta \Delta U^{(2)}} \delta(\mathbf{r}_1 - \mathbf{r})\delta(\mathbf{r}_2 - \mathbf{r}') \right\rangle\!\!\right\rangle_0}{\left\langle\!\!\left\langle e^{-\beta \Delta U^{(2)}} \right\rangle\!\!\right\rangle_0}. \tag{3.48}$$

We now restrict attention to a uniform system, insist that this pair will be treated according to classical statistical mechanics, and evaluate the average for the distinguished pair:

$$\left\langle\left\langle\delta(r_1 - r)\delta(r_2 - r')\right\rangle\right\rangle_0 = \frac{(\mathcal{Q}_1/\mathcal{V})^2}{\mathcal{Q}_2} e^{-\beta u^{(2)}(r,r')}. \tag{3.49}$$

Collecting these results permits us to express $g^{(2)}(r, r')$ in terms of evocative conditional means:

$$g^{(2)}(r, r') = e^{-\beta u^{(2)}(r,r')} \times \left[\frac{\left\langle e^{-\beta \Delta U^{(2)}} | r_1 = r, r_2 = r'\right\rangle_0}{\left\langle e^{-\beta \Delta U^{(1)}} | r_1 = r\right\rangle_0 \left\langle e^{-\beta \Delta U^{(1)}} | r_2 = r'\right\rangle_0}\right]. \tag{3.50}$$

Here $u^{(2)}(r, r')$ is the interaction potential energy for the distinguished pair at the indicated positions. If the potential energy model is not pair decomposable, $u^{(2)}(r, r')$ is the potential energy for the distinguished pair *without* the rest of the system, because it derives from the relation Eq. (3.49).

We note that the potential of mean force between the distinguished pair of atoms, given by $\beta W(r, r') \equiv -\ln g^{(2)}(r, r')$, is obtained directly from Eq. (3.50):

$$\beta W(r, r') = \beta u^{(2)}(r, r') - \ln \left[\frac{\left\langle e^{-\beta \Delta U^{(2)}} | r_1 = r, r_2 = r'\right\rangle_0}{\left\langle e^{-\beta \Delta U^{(1)}} | r_1 = r\right\rangle_0 \left\langle e^{-\beta \Delta U^{(1)}} | r_2 = r'\right\rangle_0}\right]. \tag{3.51}$$

The first term on the right side of this equation is the contribution from direct intermolecular interactions between the distinguished pair of atoms, and the second term is the contribution corresponding to indirect interactions through the solvent. The solvent-mediated interactions are computed as an average over the thermal motion of the solvent with the two distinguished atoms placed at (r, r'), relative to a second average over the uncoupled thermal motions of the solvent and each individual molecule of this distinguished pair, in each case with no coupling between the solute–solvent subsystems. This second average, which appears in the denominator or Eq. (3.51), accounts for the loss of spatial correlations as $|r - r'| \to \infty$; i.e.,

$$\lim_{|r-r'|\to\infty} g^{(2)}(r, r') = 1. \tag{3.52}$$

The p-particle joint density and associated correlation function for a one-component simple fluid follows directly from a generalization of the preceding development leading to Eq. (3.50). Thus,

$$\rho^{(p)}(r_1, \ldots, r_p) = \rho^p g^{(p)}(r_1, \ldots, r_p), \tag{3.53}$$

where

$$g^{(p)}(\boldsymbol{r}_1, \ldots, \boldsymbol{r}_p) = e^{-\beta u^{(p)}(r_1, \ldots, r_p)} \left[\frac{\left\langle e^{-\beta \Delta U^{(p)}} \middle| \boldsymbol{r}_1 \ldots \boldsymbol{r}_p \right\rangle_0}{\prod_{j=1}^{p} \left\langle e^{-\beta \Delta U^{(1)}} \middle| \boldsymbol{r}_j \right\rangle_0} \right]. \tag{3.54}$$

Exercises

3.16 Derive Eq. (3.45).

3.17 Figure 3.3 shows the potential of mean force (PMF) as a function of separation between two methane-sized cavities in water at 298 K. The PMF is normalized to a value of zero at infinite separation. The negative values that are obtained as the two cavities approach one another reflect hydrophobic driving forces that show a tendency to aggregate in water, even in the absence of interactions between the cavities. Since the cavities do not interact, they can overlap to form a single methane-sized cavity, thereby minimizing their contact with water. If, instead, two methane molecules aggregated to form a dimer in water, the methane–methane separation would on average correspond to a minimum in the methane–methane PMF. Calculate this separation assuming that the pair potential energy of interaction between two methane

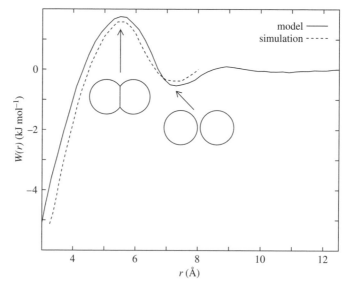

Figure 3.3 Potential of mean force between two methane-sized cavities as a function of their separation generated from molecular simulations. See Hummer *et al.* (1996).

molecules can be described by the Lennard–Jones potential function with $\sigma = 3.730\,\text{Å}$ and $\varepsilon = 0.294\,\text{kcal mol}^{-1}$. Sketch this PMF as a function of methane–methane separation.

3.7 Activity coefficients and solution standard states

In this section, we derive PDT expressions for activity coefficients and standard state chemical potentials that are conventional in physical chemistry and chemical engineering thermodynamics. We assume here a single homogeneous solution phase composed of several components, and write the following conventional expression for the chemical potential of component α in this multicomponent solution:

$$\beta\mu_\alpha = \beta\mu_\alpha^\circ + \ln\gamma_\alpha x_\alpha, \tag{3.55}$$

where μ_α° is a reference chemical potential, x_α is the mole fraction of component α, and γ_α is the activity coefficient which characterizes deviations from the reference chemical potential, μ_α°. We seek the connection between this expression and the PDT formula for this chemical potential,

$$\beta\mu_\alpha = \ln\left[\rho_\alpha\Lambda_\alpha^3/q_\alpha^{\text{int}}\right] - \ln\left\langle\!\left\langle e^{-\beta\Delta U_\alpha}\right\rangle\!\right\rangle_0. \tag{3.56}$$

Recognizing that $\rho_\alpha \equiv x_\alpha\rho$ where ρ is the total number density of all molecules in the volume $\mathcal{V}$, we can rewrite Eq. (3.56) as

$$\beta\mu_\alpha = \ln\left[\frac{\rho\Lambda_\alpha^3}{q_\alpha^{\text{int}}\left\langle\!\left\langle e^{-\beta\Delta U_\alpha}\right\rangle\!\right\rangle_0}\right] + \ln x_\alpha. \tag{3.57}$$

The first term on the right side of this equation must be equal to $\beta\mu_\alpha^\circ + \ln\gamma_\alpha$ from Eq. (3.55). Since μ_α° and γ_α are defined with respect to one another, we can multiply and divide the term in brackets by a factor that defines the standard state, and then separate the numerator and denominator into factors $\beta\mu_\alpha^\circ$ and $\ln\gamma_\alpha$.

Our first choice for a standard state is the pure fluid, and therefore, the factor of interest is the average Boltzmann factor for coupling a distinguished molecule to its pure fluid:

$$\left[\left\langle\!\left\langle e^{-\beta\Delta U_\alpha}\right\rangle\!\right\rangle_0\right]_{x_\alpha=1}. \tag{3.58}$$

Multiplying and dividing by this factor appropriately in Eq. (3.57) gives

$$\beta\mu_\alpha = \ln\left[\frac{\rho\Lambda_\alpha^3}{q_\alpha^{\text{int}}\left[\left\langle\!\left\langle e^{-\beta\Delta U_\alpha}\right\rangle\!\right\rangle_0\right]_{x_\alpha=1}}\right] + \ln\gamma_\alpha x_\alpha, \tag{3.59}$$

with the activity coefficient given by

$$\gamma_\alpha = \frac{\left[\langle\langle e^{-\beta\Delta U_\alpha}\rangle\rangle_0\right]_{x_\alpha=1}}{\langle\langle e^{-\beta\Delta U_\alpha}\rangle\rangle_0}. \tag{3.60}$$

The Boltzmann factor in the denominator of this equation corresponds to coupling a distinguished molecule of component α to the solution. This result is reminiscent of local composition free energy models that are widely used to calculate fluid-phase equilibria for multicomponent mixtures of nonelectrolytes. We note that $\gamma_\alpha > 1$ corresponds to less favorable interactions in the mixtures, and $\gamma_\alpha \to 1$ as $x_\alpha \to 1$.

Our second choice for a standard state is a component at infinite dilution in solution. In this case, we multiply and divide Eq. (3.57) by the PDT factor corresponding to the $x_\alpha = 0$ circumstance:

$$\left[\langle\langle e^{-\beta\Delta U_\alpha}\rangle\rangle_0\right]_{x_\alpha=0}, \tag{3.61}$$

and write the chemical potential as

$$\beta\mu_\alpha = \ln\left[\frac{\rho\Lambda_\alpha^3}{q_\alpha^{\text{int}}\left[\langle\langle e^{-\beta\Delta U_\alpha}\rangle\rangle_0\right]_{x_\alpha=0}}\right] + \ln\gamma_\alpha x_\alpha. \tag{3.62}$$

The first term on the right side of this equation is $\beta\mu_\alpha^\infty$, corresponding to the infinite-dilution chemical potential of component α, and the activity coefficient is defined as

$$\gamma_\alpha = \frac{\left[\langle\langle e^{-\beta\Delta U_\alpha}\rangle\rangle_0\right]_{x_\alpha=0}}{\langle\langle e^{-\beta\Delta U_\alpha}\rangle\rangle_0} = \gamma_\alpha(\{x_1, \ldots, x_n\}). \tag{3.63}$$

We note that the standard state, infinite-dilution chemical potential, μ_α^∞, involves only solute–solvent interactions, since the coupling energy in the Boltzmann factor in Eq (3.61) is computed with no other solute molecules present. If finite solute concentrations lead to more favorable Boltzmann factors, then $\gamma < 1$, and vice versa for less favorable interactions. Also, as $x_\alpha \to 0$, $\gamma_\alpha \to 1$, as expected.

Exercise

3.18 Use the expression derived above for μ_α^∞ to obtain the following microscopic expression for the Henry's law coefficient:

$$k_\alpha = \frac{\rho k T}{\left[\langle\langle e^{-\beta\Delta U_\alpha}\rangle\rangle_0\right]_{x_\alpha=0}}. \tag{3.64}$$

Describe the behavior of this coefficient in the limits of high and low solubility.

3.8 Quantum mechanical ingredients and generalizations

Classical statistical mechanical theory is, for the most part, adequate for the solutions treated in this book (Benmore *et al.*, 2001; Tomberli *et al.*, 2001), as has been discussed more specifically elsewhere (Feynman and Hibbs, 1965). It is important to distinguish that issue of statistical mechanical theory from the theory, computation, and modeling involved in the interaction potential energy $U(\mathcal{N})$. The potential distribution theorem doesn't require specifically simplified forms for $U(\mathcal{N})$ on grounds of statistical mechanical principal; simplifications can make calculations more practical, of course, but those are issues to be addressed for specific cases.

Vibrational motions of molecules are typically significantly quantized; these motions affect the q_α^{int} of Eq. (3.18), p. 40. Some problems suggest a broader quantum description than the treatment of the internal vibrations only, even though quantum mechanical effects might be secondary to the classical description. For aqueous solutions, the study of H/D isotope effects are legion. Most of the equilibrium properties of liquid water exhibit minor quantum effects, with H_2O being a slightly more disordered liquid than D_2O (Buono *et al.*, 1991; Tomberli *et al.*, 2000). That additional disorder is often described by analogy with an increase in temperature (Landau *et al.*, 1980, Section 33. Expansion in powers of $\hbar$). For liquid thermodynamic states near the triple point, the increase can be characterized by saying that the magnitude is comparable to a 5–6 K temperature rise (Buono *et al.*, 1991; Tomberli *et al.*, 2000; Badyal *et al.*, 2002). Not coincidentally, the temperature of maximum density of liquid H_2O at normal pressure (4° C) is 7° C below the corresponding temperature of maximum density of liquid D_2O, and the triple temperature of H_2O is about 4 K lower than that of D_2O. But the critical temperature of H_2O is about 3 K *higher* than the critical temperature of D_2O (Eisenberg and Kauzmann, 1969). The solubilities of nonpolar gases in H_2O at room temperature and low pressure are measurably different from D_2O, and those differences can be correlated with the differences in the compressibilities of the two solvents (Hummer *et al.*, 2000). Also, the static dielectric constant of H_2O is slightly higher than that of D_2O for temperatures not too high (Stillinger, 1982). So, despite being small, these effects are not perfectly understood. Other equilibrium properties might have higher sensitivity to these quantum effects than do the chemical potentials (Guillot and Guissani, 1998a; b).

The most basic point here is that the only specifically classical feature of Eq. (3.17), p. 40, is the *n*! assumed in Eq. (3.14), p. 39. This feature derives from the indistinguishability of particles other than the electrons, and a more correct account of the indistinguishability of those heavy particles would involve exchanging identities with the proper phases (Feynman and Hibbs, 1965). But, as is well known, those exchange contributions are the least significant of quantum mechanical effects for the solutions of interest here (Feynman and Hibbs, 1965). Other quantum mechanical effects can be described by Eq. (3.17), even if that requires a somewhat specialized treatment.

This point can be underscored by returning again to consider the unnormalized density matrix lurking underneath Eq. (2.15), p. 26:

$$\frac{\partial \rho \left(\mathcal{N}, \mathcal{N}' \right)}{\partial \beta} = -\mathbb{H} \rho \left(\mathcal{N}, \mathcal{N}' \right).$$ (3.65)

Eq. (2.15) is then translated to

$$e^{-\beta \mathcal{A}} = Tr \, \rho.$$ (3.66)

The point to be underscored is that many available approximate solutions for $\rho \left(\mathcal{N}, \mathcal{N}' \right)$ of Eq. (3.65), e.g. (Gomez and Pratt, 1998), with the Boltzmann–Gibbs treatment of heavy-particle exchange, can be applied to evaluation of the potential distribution theorem. The most important physical requirement is that such a model gracefully adapt to the classical limit, because that is the most important physical limit for molecular solutions.

In this section we discuss quantum mechanical models that can be brought to bear on evaluation of the potential distribution theorem. These models could be tried and tested in practical calculations, but the basics of these models should be studied elsewhere – the present discussion is not about quantum mechanics for its own sake. The remainder of this section then gives a more technical discussion of current ideas for inclusion of nonexchange quantum mechanical effects.

Initial treatment of quantum statistical mechanics

As a preliminary point, we note that the decoupled averaging discussed here in classical views of the potential distribution theorem derives from the denominator of Eq. (3.17), p. 40. This is unchanged in the present quantum mechanical discussion, and thus the sampling of the separated subsystems could be highly quantum mechanical without changing those formalities.

For simplicity, however, let's adopt an atomic-level description of the molecular solution. As with the central-force models (Lemberg and Stillinger, 1975), the interactions treated could be sufficiently complicated as to form complex molecules, but we will focus on the interaction contributions to the chemical potentials of atoms. In that case, all $q_\alpha^{\text{int}} = 1$, and any quantum mechanical effects on the internal structures of molecules that may be formed will have to be described by the quantum mechanical approximations that are the target here.

We discuss initially the model (Feynman and Hibbs, 1965, Section 10-3)

$$\bar{U}(\mathcal{N}) \approx U(\mathcal{N}) + \frac{\beta \hbar^2}{24} \sum_j \frac{\alpha}{m_j} \nabla_j^2 U(\mathcal{N}). \tag{3.67}$$

Here $U(\mathcal{N})$ is the interaction potential energy for the complete system at a specific configuration, uniformly the same quantity that has been discussed above. $\bar{U}(\mathcal{N})$ is an *effective potential* designed to be used in classical-limit partition function calculations, e.g. Eq. (3.17), p. 40, in order to include quantum mechanical effects approximately. We will call this $\bar{U}(\mathcal{N})$ the quadratic Feynman–Hibbs (QFH) model. In Eq. (3.67), m_j is the mass of atom j, and ∇_j^2 is the Laplacian of the cartesian positional coordinates of atom j, $\nabla_j^2 = \left(\partial/\partial x_j \right)^2 + \left(\partial/\partial y_j \right)^2 + \left(\partial/\partial z_j \right)^2$. The rightmost contribution in Eq. (3.67) becomes smaller if β decreases (higher temperature), or as the masses m_j become larger (perhaps by substituting D for H).

Because of the linear dependence of $\bar{U}(\mathcal{N})$ on $U(\mathcal{N})$, we can straightforwardly define

$$\Delta \bar{U}_\alpha^{(1)} \equiv \bar{U}(\mathcal{N}+1) - \bar{U}(\mathcal{N}) - \bar{U}(1). \tag{3.68}$$

This application of the QFH model is then

$$\beta \mu_\alpha^{\text{ex}} = -\ln \left\langle \left\langle e^{-\beta \Delta \bar{U}_\alpha^{(1)}} \right\rangle \right\rangle_0. \tag{3.69}$$

The indicated averaging is the classical averaging for the decoupled subsystems with the effective interactions $\bar{U}$ for each case. This is a remarkably simple result, and study of this model is a serious first step in understanding quantum mechanical effects in molecular solutions.

A more ambitious model than QFH, and one that is expected to be more accurate, is the Feynman–Hibbs (FH) model (Feynman and Hibbs, 1965) for which

$$\bar{U}(\mathcal{N}) \approx \int \cdots \int U(\mathcal{N} + \{x_1, x_2, \ldots\}) \prod_{\text{atoms } j} \sqrt{\frac{12 m_j}{2 \pi \beta \hbar^2}} e^{-12 x_j^2 m_j / 2 \beta \hbar^2} \, dx_j. \tag{3.70}$$

The $\{x_1, x_2, \ldots\}$ are displacements of the atoms, and this is a gaussian convolution of the mechanical potential energy $U(\mathcal{N})$ with the variances depending on the masses of the atoms, on the temperature, and on $\hbar$. If $U(\mathcal{N})$ is weakly dependent on the displacements $\{x_1, x_2, \ldots\}$, then this FH model reduces to the QFH model, Eq. (3.67). Near minima of $U(\mathcal{N})$ this gaussian convolution means that $\bar{U}(\mathcal{N}) \geq U(\mathcal{N})$; this is an approximate description of zero point motion. Near maxima of $U(\mathcal{N})$ this gaussian convolution means that $\bar{U}(\mathcal{N}) \leq U(\mathcal{N})$; this is an approximate description of barrier tunneling. Simulations of liquid water have been conducted on the basis of this Feynman–Hibbs model, with encouraging results (Guillot and Guissani, 1998*a*; *b*). Again, Eq. (3.68) provides a straightforward definition of $\Delta \bar{U}^{(1)}$, and Eq. (3.69) provides an approximate evaluation of that desired thermodynamic quantity within this model.

This discussion has adopted an atomic-level description. But we can now reconsider the typical molecular case where intramolecular vibrations are strongly quantized. The natural idea is to evaluate the isolated molecular partition functions, q_α^{int}, by standard methods, beginning with the harmonic approximation. The intermolecular, excess contribution to those chemical potentials might be evaluated by one of the simple models above, either QFH or FH. In that procedure, the sampling would be with the decoupled systems obeying the simple model chosen.

The quantum potential distribution theorem

The discussion above treated secondary quantum effects in molecular solutions. Now we examine quantum effects more fully, though still neglecting exchange effects. Our purpose is twofold. First, there are systems such as the solvated electron where a full quantum treatment is required (Marchi *et al.*, 1988). Second, the models above can be derived. We will see again that the PDT gives a particularly simple means of obtaining the approximate results.

The Feynman path integral picture of quantum mechanics is a formulation alongside the Heisenberg or Schrödinger approaches (neglecting spin) (Feynman and Hibbs, 1965; Feynman, 1972). We do not give details of the path integral method (Doll *et al.*, 1990), but rather discuss its basic features and present the results necessary to obtain the chemical potential. In the path integral formulation of equilibrium statistical mechanics, each atom is represented by a cyclic path. The representation is analogous to a cyclic polymer molecule. Here we employ a Fourier representation of the paths (Doll *et al.*, 1990) for convenience. Then the path is given by

$$x_\tau = x_0 + \sum_{k=1}^{\infty} a_k \sin k\pi\tau. \tag{3.71}$$

x_0 is the origin of the path corresponding to $\tau = 0$, or 1, and a_k is the kth Fourier variable. τ describes the evolution along the path and progresses from 0 to 1. Then the quantum potential distribution theorem (Beck, 1992; Beck and Marchioro, 1993; Wang *et al.*, 1997; Beck, 2006) for a quantum solute with no internal structure is

$$\beta\mu_\alpha = \ln \rho_\alpha \Lambda_\alpha^3 - \ln \left\langle \left\langle e^{-\beta \int_0^1 \Delta U_\alpha(x_\tau) d\tau} \right\rangle_{a_k} \right\rangle_0. \tag{3.72}$$

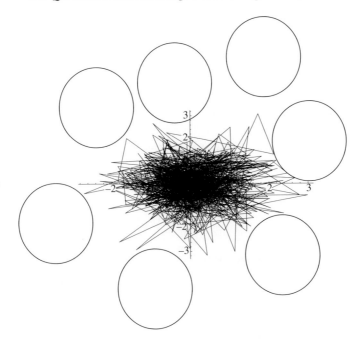

Figure 3.4 Schematic representation of a quantum particle solvated in a bath of classical molecules. The cyclic path represents the quantum particle in the field created by the classical solvent molecules.

See Fig. 3.4 for a pictorial representation of the potential distribution theorem. The inner-averaging weight is a gaussian – corresponding to the free-particle kinetic energy and specified by the path, with no coupling to the solvent particles:

$$\langle\ldots\rangle_{a_k} = \frac{\int da_1 \cdots da_k\, e^{-\sum_{k=1}^{\infty} a_k^2/2\sigma_k^2}(\ldots)}{\prod_{k=1}^{\infty} \int da_k\, e^{-a_k^2/2\sigma_k^2}}, \tag{3.73}$$

where

$$\sigma_k^2 = \frac{2\beta\hbar^2}{m\pi^2 k^2}. \tag{3.74}$$

Notice first that the formula for the chemical potential is very close to the molecular formula from the classical potential distribution theorem. The inner average is over the free-particle kinetic energy of the path, instead of over the thermal motions of the molecule in the gas phase. In either case, the solute is decoupled from the solvent. Also, instead of the interaction energy of the classical point particle with the solvent, here one needs the average interaction energy along the path. The classical limit is recovered as the temperature or the mass gets large, since the gaussian kinetic energy distributions become very narrow, and the paths shrink down to the point x_0. Then the classical potential distribution theorem for an atomic solute emerges directly from Eq. (3.72). Here we explore a re-expression of the quantum potential distribution theorem and approximations based on it which lead to the QFH and FH approximations.

The first step we take is to multiply and divide by the classical average Boltzmann factor in the potential distribution theorem:

$$\beta\mu_\alpha = \ln\rho_\alpha\Lambda_\alpha^3 - \ln\left\langle e^{-\beta\Delta U_\alpha(\bar{x})}\right\rangle_0$$

$$- \ln\frac{\left\langle\left\langle e^{-\beta\int_0^1 \Delta U_\alpha(x_\tau)d\tau}\right\rangle_{a_k}\right\rangle_0}{\left\langle\left\langle e^{-\beta\Delta U_\alpha(\bar{x})}\right\rangle_{a_k}\right\rangle_0}. \qquad (3.75)$$

The Fourier coefficient average in the denominator of the last term is included to make the numerator and denominator symmetrical; it has no effect on the classical average. Let's assume that the classical potential is evaluated at the *centroid* of the path:

$$\bar{x} = \int_0^1 x_\tau d\tau = x_0 + \frac{1}{\pi}\sum_{k=1}^\infty \frac{a_k}{k}\left[1 - (-1)^k\right]. \qquad (3.76)$$

This will prove useful in deriving the approximations. Then we can write the quantum potential distribution theorem as

$$\beta\mu_\alpha = \ln\rho_\alpha\Lambda_\alpha^3 - \ln\left\langle e^{-\beta\Delta U_\alpha(\bar{x})}\right\rangle_0$$

$$- \ln\left\langle\left\langle e^{-\beta\int_0^1 [\Delta U_\alpha(x_\tau) - \Delta U_\alpha(\bar{x})]d\tau}\right\rangle_{a_k}\right\rangle_r, \qquad (3.77)$$

where the first two terms on the right side yield the classical chemical potential, and the third term is an *exact* quantum correction. The averaging process on the correction term is over all solvent degrees of freedom with this *classical* solute included. In this way the quantum effects are computed by *reference* to the classical solute.

Now focus on the quantum correction term. We make an approximation to the correction based on the inequality (Feynman, 1972)

$$\left\langle e^{-f}\right\rangle \geq e^{-\langle f\rangle}. \qquad (3.78)$$

We make this approximation for the inner gaussian kinetic energy average. This inequality shows that our approximation will yield an upper bound to the exact chemical potential. The exercises to follow derive the Feynman–Hibbs and QFH models from this approximation.

Exercises

3.19 Simplify the inner multidimensional gaussian integral of the last term in Eq. (3.77), following the approximation Eq. (3.78), by showing that

$$
\left\langle \int_0^1 [\Delta U_\alpha (x_\tau) - \Delta U_\alpha (\bar{x})] \, d\tau \right\rangle_{a_k}
$$

$$
\equiv \frac{\int \dots \int \left\{ \int_0^1 [\Delta U_\alpha (x_\tau) - \Delta U_\alpha (\bar{x})] \, d\tau \right\} \prod_{k=1}^\infty e^{-\frac{a_k^2}{2\sigma_k^2}} \, da_k}{\prod_{k=1}^\infty \int e^{-\frac{a_k^2}{2\sigma_k^2}} \, da_k} \tag{3.79}
$$

can be expressed as

$$
\frac{\int \dots \int [\Delta U_\alpha (\bar{x} - y) - \Delta U_\alpha (\bar{x})] \prod_{k=1}^\infty e^{-\frac{a_k^2}{2\sigma_k^2}} \, da_k}{\prod_{k=1}^\infty \int e^{-\frac{a_k^2}{2\sigma_k^2}} \, da_k}, \tag{3.80}
$$

with

$$
y = \frac{1}{\pi} \sum_{k=1}^\infty \frac{a_k}{k} \left[1 - (-1)^k \right]. \tag{3.81}
$$

Hint: notice that the $\tilde{\imath}$-integration of the numerator could be performed last, and that no point along the path is special. Since the u-integral is then irrelevant, choose the path point x_0 as representative, and let y be the displacement according to $x_0 = \bar{x} - y$.

Further simplify these integrals by inserting

$$
\int \delta \left(y - \frac{1}{\pi} \sum_{k=1}^\infty \frac{a_k}{k} \left[1 - (-1)^k \right] \right) dy = 1 \tag{3.82}
$$

or

$$
\int \left\{ \int e^{i\omega \left(y - \frac{1}{\pi} \sum_{k=1}^\infty \frac{a_k}{k} [1 - (-1)^k] \right)} \frac{d\omega}{2\pi} \right\} dy = 1 \tag{3.83}
$$

into the numerator and denominator of the gaussian average, and derive the PDT analog of the Feynman–Hibbs potential of Eq. (3.70). Hint: rearrange the order of integration to do the a_k integrals first, followed by the ω integral, and note the sum

$$
\frac{1}{4} \sum_{k=1}^\infty \frac{1}{k^4} \left[1 - (-1)^k \right]^2 = \frac{\pi^4}{96}. \tag{3.84}
$$

3.20 Obtain the QFH correction from the Feynman–Hibbs effective potential by Taylor expansion of the potential.

3.21 Show that the classical chemical potential is always lower than the exact quantum chemical potential. Hint: see Predescu (2003). Therefore, the exact quantum chemical potential lies between the classical chemical potential and that obtained from the Feynman–Hibbs approximation. (Note the error in Feynman [1972, p.64] concerning this point.)

4

Models

In this chapter we develop simple models that treat characteristic problems that arise in describing solutions at a molecular level. These are models that might be used in practical considerations of molecular solutions. We also use these accumulating examples to support the view that the potential distribution theorem provides a simple, effective basis for further development of the theory of molecular liquids.

Beyond the use of the potential distribution theorem, the developments here would be described as *physical*. The goal is to discuss simple models that might be useful. More sensitive refinement can come later.

In discussing these models, we will assume that simulation data are typically available; we expect that to be the most common case. Helpful theoretical models can then be built *on top of* simulation (or experimental) data.

4.1 Van der Waals model of dense liquids

A central theme of the modern theory of liquids is a reappreciation of the van der Waals equation of state. The traditional presentations of the van der Waals equation of state feature discussion of two concepts (Widom, 2002, see Section 7.2): (*i*) a *free volume* modification of the ideal gas equation of state based on the fact that molecules can't overlap much, and (*ii*) modification of that free volume equation of state to reflect attractive interactions between molecules. The result is

$$p = \frac{\rho kT}{1 - b\rho} - a\rho^2. \tag{4.1}$$

The traditional van der Waals equation of state is a formula with two empirical parameters, a and b, one for each of the physical arguments put forward here.

The reappreciation of the van der Waals theory refines these two arguments. A decisive feature is a sharp distinction between the differing roles of attractive and repulsive interactions; the fact that these interactions of differing physical character can be treated in two steps with the additive consequences seen in Eq. (4.1) is already the most important point. For the present purposes, a van der Waals view of a real solution is this sequential reasoning: first repulsive forces and subsequently the effects of attractive interactions.

We then consider this hint from the vantage of the potential distribution theorem, Eq. (3.5). We suppose that we can separate the interaction ΔU_α into two contributions $\Delta \tilde{U}_\alpha + \Phi_\alpha$. If Φ_α were not present, we would have

$$e^{-\beta \tilde{\mu}_\alpha^{ex}} = \left\langle\!\!\left\langle e^{-\beta \Delta \tilde{U}_\alpha} \right\rangle\!\!\right\rangle_0 . \tag{4.2}$$

The tilde over $\tilde{\mu}_\alpha^{ex}$ indicates that this is the interaction contribution to the chemical potential of the solute when $\Phi_\alpha = 0$. Even though we are imagining manipulating the interactions of the solution and the distinguished solute, the properties of the solution alone are unchanged.

With the results Eq. (4.2) available, we next consider the remainder:

$$e^{-\beta(\mu_\alpha^{ex} - \tilde{\mu}_\alpha^{ex})} = \frac{\left\langle\!\!\left\langle e^{-\beta \Delta U_\alpha} \right\rangle\!\!\right\rangle_0}{\left\langle\!\!\left\langle e^{-\beta \Delta \tilde{U}_\alpha} \right\rangle\!\!\right\rangle_0} . \tag{4.3}$$

The ratio on the right side here is suggestively similar to the ratio on the right side of Eq. (3.24), p. 42, because $e^{-\beta \Delta U_\alpha} = e^{-\beta \Delta \tilde{U}_\alpha} \times e^{-\beta \Phi_\alpha}$. Consulting that previous formula, we can make the correspondence that $F = e^{-\beta \Phi_\alpha}$ and then

$$e^{-\beta(\mu_\alpha^{ex} - \tilde{\mu}_\alpha^{ex})} = \left\langle\!\!\left\langle e^{-\beta \Phi_\alpha} \right\rangle\!\!\right\rangle_r . \tag{4.4}$$

Here the $\langle\!\langle \ldots \rangle\!\rangle_r$ indicates an averaging for the case that the solution contains a distinguished molecule which interacts with the rest of the system on the basis of the function $\Delta \tilde{U}_\alpha$; the subscript r stands for *reference*.

Equation (4.4) will be broadly useful; it describes the effects of Φ_α and the physical perspective that we follow here is that the van der Waals approach treats the interactions Φ_α as a perturbation. To analyze this further we introduce the probability density function

$$\tilde{\mathcal{P}}_\alpha(\varepsilon) \equiv \langle\!\langle \delta(\varepsilon - \Phi_\alpha) \rangle\!\rangle_r . \tag{4.5}$$

The van der Waals approximation is built on the idea that Φ_α is a background molecular field that scarcely fluctuates during the course of the thermal motion of the system composed of the distinguished solute and solution. In that situation, $\tilde{\mathcal{P}}_\alpha(\varepsilon)$ is so tightly concentrated that the only significant parameter is its location, and that suggests $\tilde{\mathcal{P}}_\alpha(\varepsilon) \approx \delta(\varepsilon - \langle\!\langle \Phi_\alpha \rangle\!\rangle_r)$. Following this physical reasoning, Eq. (4.4) reduces to

$$\mu_\alpha^{ex} \approx \tilde{\mu}_\alpha^{ex} + \langle\!\langle \Phi_\alpha \rangle\!\rangle_r . \tag{4.6}$$

We will later consider the approximation that affects the transition from Eq. (4.4) to Eq. (4.6) in detail. But this result would often be referred to as *first-order perturbation theory* for the effects of Φ_α – see Section 5.3, p. 105 – and we will sometimes refer to this result as the *van der Waals approximation*. The additivity of the two contributions of Eq. (4.1) is consistent with this form, in view of the thermodynamic relation $\rho d\mu = dp$ (constant T). It may be worthwhile to reconsider Exercise 3.5, p. 39. The nominal temperature independence of the last term of Eq. (4.6), is also suggestive. Notice, however, that the last term of Eq. (4.6), as an approximate correction to $\tilde{\mu}_\alpha^{\text{ex}}$, will depend on temperature in the general case. This temperature dependence arises generally because the averaging $\langle\langle\ldots\rangle\rangle_r$ will imply some temperature dependence. Note also that the density of the solution medium is the actual physical density associated with full interactions between all particles with the exception of the sole distinguished molecule. That solution density will typically depend on temperature at fixed pressure and composition.

The restriction to dense liquids and, in some respects, to one-component systems deserves further comment. Such restrictions follow from the physical assumption here that thermal fluctuations in Φ_α are negligible. If a one-component liquid is sufficiently dense, then variety in the structures that occur with reasonable frequency during the thermal motion will be limited. Since density fluctuations may be gauged with the compressibility $\partial\rho/\partial\beta p|_T$, see Eq. (2.24) p. 27, these approximations are expected to be better where the compressibility is (or more generally, susceptibilities are) smaller. The compressibility is likely to be smaller for the higher density fluid states.

Another physical consideration is the spatial range of the interactions Φ_α. If a large number of distinct solution elements make small contributions to Φ_α, it is reasonable to hope that fluctuations in the net result would be less for that statistical reason. In typical physical cases, this statistical point is insufficient as a sole argument, but it is typically a helpful additional point.

For specific cases, it is nontrivial to identify the perturbation Φ_α that would make this simple approximation effective. Φ_α need not be pair decomposable. Furthermore, for molecules of general interest, presenting interactions of several types, there is no consensus on specifically how to separate ΔU_α into two contributions that would make this van der Waals approximation compelling; see Chen *et al.* (2005). The most specific utility of this van der Waals treatment is probably the following: if interactions of some specific type have been neglected for simplicity, a rough estimate of the error might be obtained with this first-order perturbation theory.

For atomic liquids, in contrast, that situation is different. Specific proposals for identification of a perturbative interaction have been well tested and the van der Waals treatment is very successful for dense one-component atomic systems. Since much has been written about that subject, we direct the reader to standard sources for more information on those achievements (Widom, 1967; Lebowitz and Waisman, 1980; Chandler *et al.*, 1983). Much of the authority of the van der Waals treatment stems from the detailed checking of those approximations for atomic liquids. Additionally, van der Waals treatments do have the great virtue of simplicity.

Finally, we return to consider the description of the reference case for which the interactions between the solution and the distinguished solute conform to $\Phi_\alpha = 0$. If the perturbation interactions are weaker and longer-ranged than $\Delta \tilde{U}_\alpha$, on a physical basis, that typically leaves excluded-volume (or overlap) features of intermolecular interactions to be contained in $\Delta \tilde{U}_\alpha$. The simplest model for such excluded-volume interactions is a hard-core model: configurations overlapping the van der Waals volumes of solution molecules with the distinguished solute are assigned infinitely unfavorable energies. Hard-core models and the packing problems defined in this way are significant challenges for molecular theories, despite the drastic simplification. We will return to this problem in Section 4.3, p. 73, but note here only that the free-volume feature of Eq. (4.1) is an approximate solution of that packing problem for the present development. In the simplest examples, particularly for atomic liquids, the identification of $\Delta \tilde{U}_\alpha$ with hard-sphere models is highly refined and successful.

Gaussian extension

Here we identify a natural extension of the van der Waals theory above; this also serves to elaborate some notation that will be helpful subsequently. The van der Waals model was based upon the estimate $\tilde{\mathcal{P}}_\alpha(\varepsilon) \approx \delta(\varepsilon - \langle\langle\Phi_\alpha\rangle\rangle_r)$. With the availability of the additional information $\langle\langle\delta\Phi_\alpha^2\rangle\rangle_r$, we could form the gaussian model

$$\tilde{\mathcal{P}}_\alpha(\Phi_\alpha) \approx \frac{1}{\sqrt{2\pi\langle\langle\delta\Phi_\alpha^2\rangle\rangle_r}} \exp\left[-\frac{1}{2}\frac{\delta\Phi_\alpha^2}{\langle\langle\delta\Phi_\alpha^2\rangle\rangle_r}\right], \qquad (4.7)$$

where $\delta\Phi_\alpha = \Phi_\alpha - \langle\langle\Phi_\alpha\rangle\rangle_r$. Better, however, is to acknowledge explicitly that these averages will depend on the conformation of the solute; we therefore write the conditional expectations $\langle\Phi_\alpha|\mathcal{R}^n\rangle_r$ and $\langle\delta\Phi_\alpha^2|\mathcal{R}^n\rangle_r$. More generally, we consider the partition function $\langle e^{-\beta\Phi_\alpha}|\mathcal{R}^n\rangle_r$ conditional on the positioning of the distinguished solute. Using this information, and a gaussian model distribution, the extended theory is

$$\mu_\alpha^{ex}(\mathcal{R}^n) \approx \tilde{\mu}_\alpha^{ex}(\mathcal{R}^n) + \langle\Phi_\alpha|\mathcal{R}^n\rangle_r - \frac{\beta}{2}\langle\delta\Phi_\alpha^2|\mathcal{R}^n\rangle_r. \qquad (4.8)$$

The first two terms here constitute the van der Waals approximation as discussed above. The succeeding term is a correction that lowers this free energy. The thermodynamic excess chemical potential is then obtained by averaging the Boltzmann factor of this conditional result using the *isolated solute* distribution function $s_\alpha{}^{(0)}(\mathcal{R}^n)$.

Gaussian density fluctuation theories

The following discussion is more technical but is useful in a subsequent section. We consider again perturbative interactions and Eq. (4.4). It may sometimes happen that the perturbative interactions Φ_α are uncertainly known, but preliminary calculations can obtain conditional densities and density variances. Thus, more primitive available information might be $\langle\rho_\gamma(r)|\mathcal{R}^n\rangle_r$ and $\langle\delta\rho_\gamma(r)\delta\rho_\nu(r')|\mathcal{R}^n\rangle_r$. In the traditional theory of liquids, attention is often directed to density functional aspects of the theory, and the perturbative interactions are perfectly known as a model.

To facilitate manipulation of the densities of the solution, we introduce expectations that are also conditional on the solution densities and denote these quantities as $\langle e^{-\beta\Phi_\alpha}|\mathcal{R}^n, \{\rho_\gamma(r)\}\rangle_r$. The advantage of this formulation is that some of the averaging can be reserved for the end of the calculation.

Our problem will be further specified by the description of perturbative interactions in this notation

$$\langle\Phi_\alpha|\mathcal{R}^n, \{\rho_\gamma(r)\}\rangle_r = \sum_\gamma \int_V \varphi_{\alpha\gamma}(\mathcal{R}^n, r)\rho_\gamma(r)\mathrm{d}^3r. \tag{4.9}$$

Thus, $\varphi_{\alpha\gamma}(\mathcal{R}^n, r)$ is the perturbation potential of the solution and the distinguished solute in conformation $\mathcal{R}^n$. If Φ_α is not decomposable according to solute–solvent pairs conditional on the solute conformation $\mathcal{R}^n$, then this relation introduces an effective pair interaction as the perturbation.

To analyze the van der Waals approach further, we will consider the approximation

$$\langle e^{-\beta\Phi_\alpha}|\mathcal{R}^n, \{\rho_\gamma(r)\}\rangle_r \approx \exp\left[-\beta\langle\Phi_\alpha|\mathcal{R}^n, \{\rho_\gamma(r)\}\rangle_r\right]. \tag{4.10}$$

This is a primitive van der Waals approximation for the indicated conditional expectation. The associated physical argument is that, with the density specified, further assessment of fluctuations is less important.

The final ingredient in this theory is an assumption permitting averaging with respect to the density fluctuations to eliminate the condition here on the densities. The simplest such assumption is that these coordinates obey a gaussian functional distribution built on the information $\langle\rho_\gamma(r)|\mathcal{R}^n\rangle_r$ and $\langle\delta\rho_\gamma(r)\delta\rho_\nu(r')|\mathcal{R}^n\rangle_r$; this is a standard idea of the random-phase approximation and related theories (Brout and Carruthers, 1963). Then

$$
\left\langle e^{-\beta\Phi_\alpha}|\mathcal{R}^n\right\rangle_r \approx \exp\left[-\sum_\gamma\int_V \beta\varphi_{\alpha\gamma}(\mathcal{R}^n,r)\left\langle\rho_\gamma(r)|\mathcal{R}^n\right\rangle_r d^3r\right]
$$

$$
\times \exp\left[\frac{1}{2}\sum_{\gamma\nu}\int_V\int_V \beta\varphi_{\alpha\gamma}(\mathcal{R}^n,r)\left\langle\delta\rho_\gamma(r)\delta\rho_\nu(r')|\mathcal{R}^n\right\rangle_r \beta\varphi_{\nu\alpha}(\mathcal{R}^n,r')d^3r\,d^3r'\right].
$$

(4.11)

In evaluating this last average, see Exercise 4.1 below, for the excess chemical potential. Then

$$
\mu_\alpha^{ex}(\mathcal{R}^n) \approx \tilde{\mu}_\alpha^{ex}(\mathcal{R}^n) + \sum_\gamma\int_V \beta\varphi_{\alpha\gamma}(\mathcal{R}^n,r)\left\langle\rho_\gamma(r)|\mathcal{R}^n\right\rangle_r d^3r
$$

$$
-\frac{\beta}{2}\sum_{\gamma\nu}\int_V\int_V \beta\varphi_{\alpha\gamma}(\mathcal{R}^n,r)\left\langle\delta\rho_\gamma(r)\delta\rho_\nu(r')|\mathcal{R}^n\right\rangle_r \beta\varphi_{\nu\alpha}(\mathcal{R}^n,r')d^3r\,d^3r'. \quad (4.12)
$$

This result should be compared to Eq. (4.8), p. 64. The information $\left\langle\delta\rho_\gamma(r)\delta\rho_\nu(r')|\mathcal{R}^n\right\rangle_r$ offers three-body and higher correlation information implicitly because of the involvement of the reference interactions and the condition on molecular configuration $\mathcal{R}^n$. In the usual notation of the theory of simple liquids, without the indicated conditions, we would write

$$
\left\langle\delta\rho_\gamma(r)\delta\rho_\nu(r')\right\rangle = \rho_\gamma\delta_{\gamma\nu}\delta(r-r') + \rho_\gamma\rho_\nu\left(g_{\gamma\nu}^{(2)}(|r-r'|)-1\right). \quad (4.13)
$$

The subscript qualifier on $\langle\ldots\rangle_r$ (Eq. (4.12)) requests that these quantities be obtained for the reference system in which a distinguished solute interacts with the solution through the defined reference system interactions. This distinction will have appreciable consequences only in the locality of the distinguished solute. Thus, if $\varphi_{\alpha\gamma}(\mathcal{R}^n,r)$ contributes to these formulae predominantly at large distances from the distinguished solute, it is reasonable to anticipate substitution of Eq. (4.13) for the correlation information appearing in Eq. (4.12). This last maneuver alleviates a subtle inconsistency in this physical discussion, where we argued that Φ_α might not be perfectly known, but assumed that $\Delta\tilde{U}_\alpha$ was perfectly known.

Exercises

4.1 The discussion above emphasized the simplicity that can follow from identification of a physical reference system. Show that

$$
\beta\mu_\alpha^{ex} = -\ln\int_{-\infty}^{\bar{\varepsilon}}\mathcal{P}_\alpha^0(\varepsilon)\,d\varepsilon + \ln\int_{-\infty}^{\bar{\varepsilon}}\mathcal{P}_\alpha(\varepsilon)e^{\beta\varepsilon}d\varepsilon, \quad (4.14)
$$

which doesn't require definition of a reference system, and thus, for example, also applies to systems with non-pair-decomposable interactions. Compare this result with Eq. (4.6), p. 62, and discuss the terms involved in that comparison. Show that an analogous formula with a smooth cut-off is obtained instead of the sharp cut-off of the integrals at $\bar{\varepsilon}$.

4.2 Follow Eq. (3.25), p. 42, and determine the probability corresponding to the averaging $\langle\langle\ldots\rangle\rangle_{\mathrm{r}}$ of Eq. (4.4).

4.3 Determine the interaction contribution to the chemical potential implied by the equation of state Eq. (4.1).

4.4 Consider the simple case where the radial distribution function in the fluid is zero for radii less than a cut-off value determined by the size of the hard core of the solute, and one beyond that value. Calculate the value of the parameter a appearing in the equation of state Eq. (4.1) for a potential of the form cr^{-n}, where c is a constant and n is an integer. An example is the Lennard–Jones potential where $n = 6$ for the long-ranged attractive interaction. What happens if $n \leq 3$? Explain what happens physically to resolve this problem. See Widom (1963) for a discussion of the issue of thermodynamic consistency when constructing van der Waals and related approximations.

4.5 Consider the several random variables x_j, $j = 1, \ldots n$. Suppose that these x_j are distributed according to a multi-variable gaussian distribution with means $\langle x_j \rangle$ and covariances $\langle \delta x_j \delta x_k \rangle = \langle x_j x_k \rangle - \langle x_j \rangle \langle x_k \rangle$. Show that

$$\langle \exp\left[-\lambda_j x_j\right]\rangle = \exp\left[-\lambda_j \langle x_j \rangle\right] \exp\left[\frac{1}{2}\lambda_j \langle \delta x_j \delta x_k \rangle \lambda_k\right]. \tag{4.15}$$

Summation over repeated subscripts from $1, \ldots, n$ is implied. Compare this result with Eqs. (4.11) and (4.12) (van Kampen, 1992).

4.2 Dielectric solvation – Born – models

A virtue of the PDT approach is that it enables precise assessment of the differing consequences of intermolecular interactions of differing types. Here we use that feature to inquire into models of electrostatic interactions in biomolecular hydration. This topic provides a definite example of the issues discussed in the previous section; the perturbative interactions, Φ_α of the previous section, are just the classic electrostatic interactions.

We note that if all electrostatic interactions between the solute and solvent are annulled, for example by eliminating all solute partial charges in force-field models, the potential distribution formula sensibly describes the hydration of that hypothetical solute:

$$e^{-\beta(\mu_\alpha^{\mathrm{ex}} - \tilde{\mu}_\alpha^{\mathrm{ex}})} = \langle\langle e^{-\beta\Phi_\alpha}\rangle\rangle_{\mathrm{r}}. \tag{4.16}$$

Following the notation of the previous section, Eq. (4.2), p. 62, the tilde indicates distinguished solute–solvent interactions without electrostatic interactions. The contribution of electrostatic interactions is then isolated as

$$\mathrm{e}^{-\beta(\mu_\alpha^{\mathrm{ex}} - \tilde{\mu}_\alpha^{\mathrm{ex}})} = \int \tilde{\mathcal{P}}_\alpha(\varepsilon)\mathrm{e}^{-\beta\varepsilon}\mathrm{d}\varepsilon, \tag{4.17}$$

following Eq. (4.4), p. 62. $\tilde{\mathcal{P}}_\alpha(\varepsilon)$ is defined by Eq. (4.5), p. 62, and $\Phi_\alpha = \Delta U_\alpha - \Delta \tilde{U}_\alpha$ is the electrostatic contribution to the solute–solvent interactions. The probability distribution, $\tilde{\mathcal{P}}_\alpha(\varepsilon)$, can be modeled using available information about the system (Hummer *et al.*, 1997). Those models should be sufficiently accurate, but it is just as important that their information content be clear so that physical conclusions might be drawn from the observed accuracy.

Our consideration of this problem benefits from the availability of simple thermodynamic models for the contribution sought. In particular, a dielectric model, the *Born model*, for the hydration free energy of a spherical ion of radius R_α with a charge q_α at its center is

$$\mu_\alpha^{\mathrm{ex}} \approx \tilde{\mu}_\alpha^{\mathrm{ex}} - \frac{q_\alpha^2}{2R_\alpha}\left(\frac{\epsilon - 1}{\epsilon}\right). \tag{4.18}$$

The dielectric constant of the external medium is ϵ. For this discussion, the significant point of Eq. (4.18) is that the electrostatic contribution is proportional to q_α^2. To emphasize this point, we change variables here, writing $\varepsilon = q_\alpha\varphi$, and consider the gaussian model (Levy *et al.*, 1991; Hummer *et al.*, 1998*b*),

$$\tilde{\mathcal{P}}_\alpha(\varphi) \approx \frac{1}{\sqrt{2\pi \langle\langle\delta\varphi^2\rangle\rangle_{\mathrm{r}}}} \exp\left[-\frac{1}{2}\frac{\varphi^2}{\langle\langle\delta\varphi^2\rangle\rangle_{\mathrm{r}}}\right]. \tag{4.19}$$

Using this probability distribution produces

$$\mu_\alpha^{\mathrm{ex}} \approx \tilde{\mu}_\alpha^{\mathrm{ex}} - \frac{\beta q_\alpha^2}{2}\langle\langle\delta\varphi^2\rangle\rangle_{\mathrm{r}}, \tag{4.20}$$

which we can compare with Eq. (4.18). The subscript notation on $\langle\langle\ldots\rangle\rangle_{\mathrm{r}}$ here indicates that averaging is performed with the solute molecule present, but in the absence of solute–solvent electrostatic interactions. This formula avoids the serious issue of the parameterization required for R_α in Eq. (4.18). It is clear from Eq. (4.20) that this parameter should depend on thermodynamic state, i.e. temperature, pressure, and composition of the system. The dependences on temperature, pressure, and composition of the gaussian model free energy have not received sufficient attention. The most primitive considerations of these questions are not heartening (see Eqs. (1.10) and (1.11), pp. 11–12), but this model is simple, physical, and worthy of further development.

We here extend the adjective *realizable* (Lesieur, 1997) to mean theoretical models obtained from an admissible probability distribution in evaluating the average Eq. (4.17). Thus, use of Eq. (4.19) produces the realizable model Eq. (4.20). Accuracy in describing valid data is, of course, a further characteristic of interest. Truncation of series expansions customary to the statistical thermodynamics of solutions can produce nonrealizable results.

Potential of the phase

An improvement is to permit Eq. (4.19) to account for a nonzero mean electrostatic potential exerted by the solution on the distinguished reference solute, writing

$$\tilde{\mathcal{P}}_\alpha(\varepsilon|\mathcal{R}^n) \approx \frac{1}{\sqrt{2\pi \langle \delta\phi_\alpha^2|\mathcal{R}^n\rangle_r}} \exp\left[-\frac{1}{2}\frac{(\varepsilon - \langle\phi_\alpha|\mathcal{R}^n\rangle_r)^2}{\langle\delta\phi_\alpha^2|\mathcal{R}^n\rangle_r}\right], \tag{4.21}$$

permitting the notation to revert to the general molecular solute case with Φ_α the electrostatic interaction impressed by the solution on the distinguished solute. With this modification Eq. (4.20) becomes

$$\mu_\alpha^{ex}(\mathcal{R}^n) \approx \tilde{\mu}_\alpha^{ex}(\mathcal{R}^n) + \langle\phi_\alpha|\mathcal{R}^n\rangle_r - \frac{\beta}{2}\langle\delta\phi_\alpha^2|\mathcal{R}^n\rangle_r. \tag{4.22}$$

The thermodynamic chemical potential is then obtained by averaging the Boltzmann factor of this conditional result using the *isolated solute* distribution function $s_\alpha^{(0)}(\mathcal{R}^n)$.

Consideration of the quantity $\langle\phi_\alpha|\mathcal{R}^n\rangle_r$ requires some conceptual subtlety. This is intended to be the electrostatic potential of the solution *induced* by reference interactions between the solute and the solution. Any contribution to the electrostatic potential that exists in the absence of those reference interactions we will call the electrostatic *potential of the phase*, but of course only electrostatic potential *differences*, e.g. between uniform conducting materials, are expected to be physically interesting.

The latter point suggests several further observations. First, we are free to adopt an arbitrarily chosen value for the potential of the phase for convenience. The value zero (0) is such a choice, and a natural choice for detailed calculations. Second, if we take a linear combination of μ_α^{ex} corresponding to neutral collections of ions, then the value of the potential of the phase will contribute zero (0) to that linear combination, because the contribution would take the form $\varphi(\sum_\alpha q_\alpha) = 0$. Those neutral linear combinations of μ_α^{ex} are thermodynamically measurable. (A specific example of experimental comparisons for neutral linear combinations is shown in Fig. 8.23, p. 212.)

With two conducting fluids in coexistence, the values of the electrostatic potentials of the phases can be regarded as a mechanical property obtainable by solution of the appropriate Poisson equation of electrostatics given sufficient information on charge densities and boundary conditions. Since the mean electric fields in the interior of a conductor vanish, electrostatic potential changes can be associated with mean electric fields at the interface between the conductors considered here. The electrostatic potential changes are then sometimes called contact potentials (Landau and Lifshitz, 1975, Section 22. The contact potential; Pratt, 1992). If those interfacial electric fields were properties of the solvents only, then it would be reasonable to attempt to determine them by experiment and calculation (Oppenheim, 1964; Pratt, 1992). Alas, these potential differences are directly influenced by the nature of the ionic solutes that are present. Ionic solutes make a Donnan-like contribution to contact potentials because short-ranged solute–solvent interactions compete with electrostatic effects to determine the ultimate solubility of a salt between two phases (Zhou *et al.*, 1988; Asthagiri *et al.*, 2003*b*). This competition is nontrivial even at low concentrations, and the effects do not vanish even in the limit of infinite dilution of those solutes. This leads to the paradoxical-sounding, but true, statement that this ionic effect on the contact potential is there "even though the ions aren't" (Pratt, 1992). Then, of course, it has to be asked which ions aren't there. A less inflammatory view is that infinitesimal concentrations of ions can have a finite effect on the contact potential.

These subtleties sometimes lead to a casual view of detailed molecular calculations such as are suggested by Eq. (4.22). If the potential of the phase is always irrelevant to neutral linear combinations of μ_α^{ex} which are thermodynamically measurable, then perhaps it is unimportant to be precise about an assumed value. Our suggestion is that results obtained by molecularly detailed calculations of solvation free energies of single ions are compared and tabulated. Thus, precision and clarity in the assumptions underlying a calculated or tabulated result are important. Indeed, if calculated or tabulated values based upon different assumptions for the potential of the phase were to be combined, it would be essential that the assumptions be precisely known. Nevertheless, an ultimate thermodynamic test of a calculation should be made on thermodynamically measurable combinations of single ion free energies.

Multigaussian extensions

A number of directions can be taken to generalize these distributions systematically. Multigaussian models are natural possibilities, and suggest the quasichemical theory taken up later. Here we assume that the distribution $\tilde{P}_\alpha(\varepsilon|\mathcal{R}^n)$ in Eq. (4.19) can be expressed as a linear combination of gaussians corresponding to configurational substates of the system. As an example for aqueous solutions, the substates might be distinct configurations defined by different hydrogen bonding configurations for the solute and solvent molecules. Indexing those substates by s, we then analyze the joint probability distribution of Φ_α and s, $\tilde{\mathcal{P}}_\alpha(\varepsilon, s|\mathcal{R}^n)$, assuming the conditional probability distributions $\tilde{\mathcal{P}}_\alpha(\varepsilon|s, \mathcal{R}^n)$ to be gaussian, and that the marginal distributions $p_\alpha(s|\mathcal{R}^n) = \int \tilde{\mathcal{P}}_\alpha(\varepsilon, s|\mathcal{R}^n)\,d\varepsilon$ are available from simulation calculations. Then

$$\tilde{\mathcal{P}}_\alpha(\varepsilon|\mathcal{R}^n) = \sum_s \tilde{\mathcal{P}}_\alpha(\varepsilon|s, \mathcal{R}^n)p_\alpha(s|\mathcal{R}^n) \qquad (4.23)$$

is the total probability formula. See Fig. 4.1. This approach effectively fixes the most immediate difficulties of dielectric models (Hummer *et al.*, 1997). The total probability formula Eq. (4.23) is generally valid, and typically a helpful *divide-and-conquer* step. But two further issues then require resolution. The first issue is the definition of *substate*, s, and the second issue is the adoption of models for $\tilde{\mathcal{P}}_\alpha(\varepsilon|s, \mathcal{R}^n)$; the discussion here emphasizes gaussian models.

Hydrogen bonds formed by the solute and solvent molecules were suggested for the case of aqueous solutions as an example of a possible scheme for cataloging configurational substates. This example again hints that the partition function, Eq. (4.17), is local in nature; in this case, its evaluation relies on the local composition in the vicinity of the solute. That this local composition is an important general property in organizing configurational substates is the most basic concept for the quasi-chemical theory developed later.

Exercises

4.6 Work out the Born model (Pettitt, 2000) result for a spherical ion of radius R_α with a charge q_α at its center embedded in a dielectric continuum with dielectric constant ϵ.

4.7 Consider a polyatomic solute for which the solute–solution electrostatic interactions are described with more than one partial atomic charge. How would you generalize the approximate models of this section?

4.8 Assume that the conditional probabilities $\tilde{\mathcal{P}}_\alpha(\varepsilon|s, \mathcal{R}^n)$ are general gaussian distributions and establish the expression for μ_α^{ex} implied by the multigaussian models in terms of observed populations $p_\alpha(s|\mathcal{R}^n)$ and the parameters (means and variances) associated with the gaussian conditional distributions. Hint: see Hummer *et al.* (1997).

4.9 A common and sensible criticism of gaussian models such as Eq. (4.22) is that the integral of Eq. (4.21) is sensitive to the $\varepsilon < 0$ wings of the distribution because of the exponential weighting. But the distribution is characterized typically on the basis of thermal fluctuations associated with kT energy

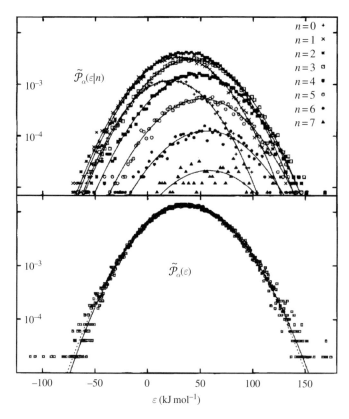

Figure 4.1 Multigaussian analysis of the probability distribution $\tilde{\mathcal{P}}_\alpha(\varepsilon)$ for an atomic ion in water ($T = 298\,\mathrm{K}$ and $\rho = 0.99707\,\mathrm{g\,cm^{-3}}$), redrawn from Hummer *et al.* (1997). The reference system is an atomic solute which interacts with water molecules through a Lennard–Jones potential model, and the independent variable ε is $|e|$ times the electrostatic potential at the center of the reference atom. For positive ions, the left wing of the graph is relevant, and the right side of the graph describes the low energy behavior in the negative ion case. The substate variable is the number of neighbor water molecules where *neighbor* is defined as a water molecule with either H atom within $3.2\,\text{Å}$ of the solute center. The symbols show histogram data. The dashed lines show the gaussian probability densities with the estimated mean and variance for each distribution. The solid line of the bottom panel is the result of combining gaussian probability densities of the conditional distributions $\tilde{\mathcal{P}}_\alpha(\varepsilon|n)$ for $n = 1$ to 7. Notice that the modes of the unconditional distributions are at positive ε, and the multigaussian model is skewed towards positive ε. This corresponds to the known phenomenon that negative ions are more strongly hydrated than positive ions. μ^{ex} is sensitive to behavior of these distributions outside the histogram data. For such reasons this example calculation is only qualitatively successful for negative ions.

scales. These different ranges might not match as a practical matter, and therefore the integral of Eq. (4.21) might not be reliably established that way. Suppose that you are able to perform a second calculation to characterize $\mathcal{P}_\alpha(\varepsilon)$, the distribution of electrostatic solute–solvent coupling for the fully coupled system. Explain why

$$e^{-\beta(\mu_\alpha^{ex} - \tilde{\mu}_\alpha^{ex})} = \frac{\int_{\bar{\varepsilon}}^\infty \tilde{\mathcal{P}}_\alpha(\varepsilon) e^{-\beta\varepsilon} \, d\varepsilon}{\int_{\bar{\varepsilon}}^\infty \mathcal{P}_\alpha(\varepsilon) \, d\varepsilon},$$ (4.24)

independently of the cut-off parameter $\bar{\varepsilon}$. Work out the form this approximation takes when each of the required distributions is modeled as a gaussian distribution with distinct parameters.

4.3 Excluded volume interactions and packing in liquids

We now consider packing issues associated with the reference term of van der Waals approaches exemplified by Eq. (4.1), p. 61. As noted there, the simplest interaction model appropriate for those terms is a *hard-core* model. The distinguished molecule considered will perfectly repel solvent molecules from an overlap, or excluded, volume. The general issues we develop will apply to such molecules in general solvents, i.e., the solvent need not be simple in the same sense as the hard-core solute we treat. But we will exemplify our general conclusions with results on the hard-sphere solvent system. The notation here, however, continues to use the tilde, as $\tilde{\mu}_\alpha^{ex}(\mathcal{R}^n)$, to indicate that the distinguished solute might serve as a reference case for subsequent treatment of other interactions.

For such models the interaction part of the chemical potential of the solute is obtained as

$$\beta\tilde{\mu}_\alpha^{ex}(\mathcal{R}^n) = -\ln p_\alpha(0|\mathcal{R}^n),$$ (4.25)

with $p_\alpha(0|\mathcal{R}^n)$ the probability that the hard-core solute could be inserted into the system without overlap of van der Waals volume of the solvent. This is a specialization of the potential distribution formula Eq. (3.5), p. 33, according to the following argument: for the hard-core solute being considered, ΔU_α is either *zero* or *infinity*, so the average sought involves a random variable with value either *one* or *zero*; the averaging collects the fraction of solute placements that would be allowed, those that score *one*. If presented with a thermal configuration of a large volume of solvent, we might estimate these quantities by performing many trial placements of the solute throughout the solvent, and determining the fraction of those trial placements that would be allowed. This estimates the fractional volume, $\mathcal{V}_{free}/\mathcal{V}$, accessible to the solute, or in other words the available volume fraction. Thus, Eq. (4.25) is a free-volume formula (Reiss, 1992), exact for the model being considered.

The operation of this formula can be viewed alternatively: imagine identifying a molecular-scale volume at an arbitrary position in the liquid system by, first, hypothetical placement of the solute, and, second, determination of those positions

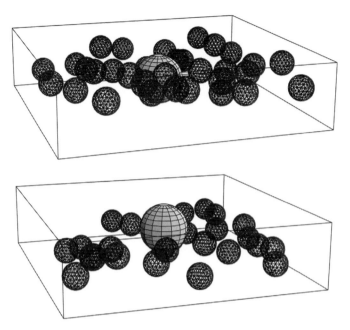

Figure 4.2 A configuration of a hard-sphere fluid sectioned to show an observation volume. The observation volume, the large sphere with radius $2d$, is shown in each section. To reconstruct the subsystem from the sections, overlap the observation ball shown in each section. This is a dense gas thermodynamic state $\rho d^3 = 0.277$.

of solvent molecules that would be excluded due to solute–solvent interactions. We will call this volume the observation volume. With such a molecular-scale volume defined, we could keep track, during a simulation calculation, of the probabilities $p_\alpha(k|\mathcal{R}^n)$ that $k = 0, 1, \ldots$ solvent-molecule occupants are observed. As the notation suggests, $p_\alpha(0|\mathcal{R}^n)$ is the probability that no occupants are observed in the molecular volume.

Our strategy for predicting $p_\alpha(0|\mathcal{R}^n)$ will be to model the distribution $p_\alpha(k|\mathcal{R}^n)$, and to extract the extreme member $p_\alpha(0|\mathcal{R}^n)$. We model the probabilities $p_\alpha(k|\mathcal{R}^n)$ on an information theory basis. We consider a relative or cross-information entropy (Shore and Johnson, 1980),

$$\eta(\{p_\alpha(k|\mathcal{R}^n)\}) = -\sum_{k=0}^{\infty} p_\alpha(k|\mathcal{R}^n) \ln\left[\frac{p_\alpha(k|\mathcal{R}^n)}{\hat{p}_\alpha(k|\mathcal{R}^n)}\right], \tag{4.26}$$

where $\hat{p}_\alpha(k|\mathcal{R}^n)$ represents a default model chosen heuristically. We anticipate that the kth-order binomial moments $\left\langle \binom{n}{k} | \mathcal{R}^n \right\rangle_0$ will be used typically. We then maximize this information entropy subject to the constraints that the probabilities reproduce the available information.

An appropriate qualitative view of such maximum-entropy modeling is that it is a *betting strategy*. More specifically, this procedure identifies an assignment of probabilities that corresponds to the sampling experiment that, relative to $\hat{p}_\alpha(j|\mathcal{R}^n)$, is maximally degenerate consistent with the constraint of the provided information. The inclusion of the default model has the consequence that this formulation generalizes gracefully as the random variable approaches a continuous limiting circumstance, permitting natural transformations of the random variable.

The formal maximization of this entropy is familiar (Jaynes, 2003). We consider the functional

$$\eta - \sum_{k=0}^{k_{\max}} \langle m^{\underline{k}}|\mathcal{R}^n\rangle_0 \frac{\zeta_k}{k!}, \tag{4.27}$$

with ζ_k being Lagrange undetermined multipliers for the moment constraints, and

$$\langle m^{\underline{k}}|\mathcal{R}^n\rangle_0 = \sum_{m\geq 0} m^{\underline{k}} p_\alpha(m|\mathcal{R}^n). \tag{4.28}$$

Here we use the *m*-to-the-*k*-falling notation of Eq. (3.42), p. 48, $m^{\underline{k}} \equiv k!\binom{m}{k}$. The requirement that this functional of the distribution $\{p_j\}$ be stationary with respect to first-order variations in the distribution $\{p_j\}$ yields

$$-\ln\left[\frac{p_\alpha(j|\mathcal{R}^n)}{\hat{p}_\alpha(j|\mathcal{R}^n)}\right] = \sum_{k=0}^{k_{\max}} j^{\underline{k}} \frac{\zeta_k}{k!}$$

$$\approx \zeta_0 + j\zeta_1 + j(j-1)\frac{\zeta_2}{2} + \cdots \tag{4.29}$$

The Lagrange multipliers ζ_k are adjusted so that the probabilities finally reproduce the information given initially. For example, ζ_0 establishes the normalization; thus

$$\sum_{k\geq 0} p_\alpha(k|\mathcal{R}^n) = 1$$

$$= e^{-\zeta_0} \sum_{m\geq 0}\left(\hat{p}_\alpha(m|\mathcal{R}^n)\prod_{k=1}^{k_{\max}} e^{-\zeta_k m^{\underline{k}}/k!}\right) \tag{4.30}$$

so that

$$e^{\zeta_0} = \sum_{m\geq 0}\left(\hat{p}_\alpha(m|\mathcal{R}^n)\prod_{k=1}^{k_{\max}} e^{-\zeta_k m^{\underline{k}}/k!}\right). \tag{4.31}$$

Therefore, using Eqs. (4.25), (4.29), and (4.31), the final thermodynamic result can be given in terms of the required normalization factor

$$\beta\tilde{\mu}_\alpha^{\mathrm{ex}}(\mathcal{R}^n) = \ln\left[1 + \sum_{m\geq 1}\left(\frac{\hat{p}_\alpha(m|\mathcal{R}^n)}{\hat{p}_\alpha(0|\mathcal{R}^n)}\prod_{k=1}^{k_{\max}} e^{-\zeta_k m^k/k!}\right)\right]. \qquad (4.32)$$

This is suggestive of the calculation of a partition function for a modest-sized set of states with effective interactions. The interactions are *n-functional* interactions, in contrast to density functional (or *ρ-functional*) theories, but with strength parameters adjusted to conform to the data available. A standard procedure for obtaining the Lagrange multipliers is to minimize the function

$$f\left(\zeta_1,\ldots,\zeta_{k_{\max}}\right) = \ln\left[1 + \sum_{m\geq 1}\left(\hat{p}_\alpha(m|\mathcal{R}^n)\prod_{k=1}^{k_{\max}} e^{-\zeta_k m^k/k!}\right)\right]$$
$$+ \sum_{k=1}^{k_{\max}}\langle m^k|\mathcal{R}^n\rangle_0 \frac{\zeta_k}{k!}. \qquad (4.33)$$

This will become operationally problematical if more Lagrange multipliers ζ_k are used than the data warrant. If operational problems are encountered, reducing $k_{\max}$ usually fixes those problems.

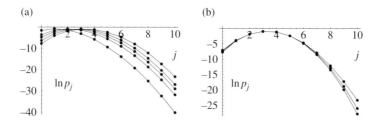

Figure 4.3 $\ln p_j$ vs. j, exemplifying the packing model of Eq. (4.29), for a hard-sphere solute in the hard-sphere fluid. Since there is no conformational flexibility in this application, we adopt the simpler notation $p_j \equiv p(j|\mathcal{R}^n)$. This example utilizes the data of Table 4.1 and $\hat{p}_j \propto 1/j!$. (a) The distributions inferred with the lowest two moments of Table 4.1 for each density of that table, low to high respectively corresponding to low to high on the right side of that panel. (b) Changes with increasing number of moments used (2 through 4) for the highest density of Table 4.1. Because $n^k = 0$ for non-negative integer $n < k$, direct effects of adding successively higher order binomial moments are seen in $\ln p_j$ only for $j \geq k$. Indirect effects come in through the normalization factor that gives the thermodynamic quantity Eq. (4.32). But since contributions with j greater than the mean are relatively small, and absolute errors typically smaller yet with inclusion of binomial moments of order k greater than the mean, convergence of the chemical potential is expected to be simple when the order of the moments used is greater than the mean.

Table 4.1 *Binomial moments* $B_k \equiv \langle \binom{n}{k} \rangle_0$ *for a spherical observation volume for unit diameter in hard-sphere fluids.* (Pratt *et al.*, 2001)

ρd^3	B_1	B_2	B_3	B_4	B_5	B_6
0.277	1.16	0.43	0.06	–	–	–
0.5	2.09	1.56	0.51	0.08	–	–
0.6	2.51	2.33	0.99	0.20	0.02	–
0.7	2.93	3.26	1.73	0.45	0.05	–
0.8	3.35	4.39	2.84	0.94	0.15	0.01

Exercises

4.10 Suppose you wish to find an extreme value of $f(x)$ on a surface $g(x) = c$, with c a constant. Explain why the gradients of f and g must be parallel at the desired extrema. Suppose you wish to find an extreme value of $f(x)$ on a surface for which $g_1(x) = c_1$ and $g_2(x) = c_2$. Explain why the gradient of f must be contained in the plane defined by the gradients of g_1 and g_2 at a desired extremum. Explain what this has to do with the Lagrange undetermined multiplier calculation.

4.11 Show how Eq. (4.29) follows from making the object Eq. (4.27) stationary with respect to variations of the probabilities.

4.12 Show that the equations $\partial f/\partial \zeta_k = 0$, $k = 1, \ldots, k_{max}$ locate the Lagrange multipliers that satisfy the moment data of Eq. (4.28). Consider the second derivatives of $f(\zeta_1, \ldots, \zeta_{k_{max}})$ to establish that the stationary point sought is a minimum.

4.13 Show that the use of only $\langle n|\mathcal{R}^n\rangle_0$ and the natural default model $\hat{p}_j \propto 1/j!$ produces the Poisson distribution and further that $\zeta_1 = -\ln\langle n|\mathcal{R}^n\rangle_0$ and $\beta\tilde{\mu}_\alpha^{ex}(\mathcal{R}^n) = \langle n|\mathcal{R}^n\rangle_0$. Introducing the notation

$$2b^{(2)}(\mathcal{R}^n)\rho \equiv \langle n|\mathcal{R}^n\rangle_0 = v\rho, \tag{4.34}$$

with v the molecular excluded volume, and $b^{(2)}(\mathcal{R}^n)$ the second virial coefficient for this conformation $\mathcal{R}^n$, then this approximation is the second virial approximation. Evaluate the pressure of the system at this level of approximation.

4.14 Consider a liquid solvent composed of one atomic type, e.g., liquid N_2, and the probability density $4\pi\lambda^2\rho\mathcal{D}_1(\lambda)$ of the distance λ to the nearest atomic center of an arbitrarily chosen point. For the case of a spherical distinguished solute, use the notation that $p_\lambda(0)$ is the insertion probability for a solute that presents a spherical excluded volume of radius λ. Show that

$$\frac{dp_\lambda(0)}{d\lambda} = -4\pi\lambda^2 \rho \mathcal{D}_1(\lambda) \tag{4.35}$$

where λ is the distance of closest approach for a solvent atom to a hard-sphere solute. Hint: explain why

$$p_\lambda(0) = 4\pi\rho \int_\lambda^\infty R^2 \mathcal{D}_1(R)\, dR. \tag{4.36}$$

4.15 Generalize the result of Eq. (4.36) by considering the distribution of the distance to the nth neighbor of an arbitrary point $4\pi\lambda^2 \rho \mathcal{D}_n(\lambda)$ and proving

$$\sum_{j=0}^{n-1} p_\lambda(j) = 4\pi\rho \int_\lambda^\infty R^2 \mathcal{D}_n(R)\, dR. \tag{4.37}$$

4.4 Flory–Huggins model of polymer mixtures

The Flory–Huggins theory is a workhorse model of phase equilibria involving polymeric fluids. It gives the description

$$\frac{\beta \Delta \mathcal{G}_{\mathrm{mix}}}{N} = x_1 \ln \phi_1 + x_2 \ln(1-\phi_1) + \mathcal{X}_{12}\phi_1(1-\phi_1) \tag{4.38}$$

of the Gibbs free energy of mixing of two pure liquids. $N = n_1 + n_2$ is the total number of molecules, x_1 is the mole fraction of component 1, and ϕ_1 is the volume fraction of that component,

$$\phi_1 = \frac{\rho_1}{\bar{\rho}_1}, \tag{4.39}$$

with ρ_1 the number density of species 1 in the mixture and $\bar{\rho}_1$ the corresponding quantity in the pure liquid. The last term of Eq. (4.38), involving the parameter $\mathcal{X}_{12}$, describes interactions between the two components. The other terms supply an approximate description of the entropy of mixing.

In recent years, there have been substantial efforts to measure the composition dependence of the interaction parameter $\mathcal{X}_{12}$ (Bates *et al.*, 1988; Han *et al.*, 1988) – see Fig. 4.4, p. 79 – and thus to refine our understanding of this theory. The original and customary derivations of the Flory–Huggins model introduced an interaction parameter that was considered to be independent of ϕ_1, but it has long been recognized that this is not the typical case (Flory, 1970). We depart here from the customary derivations to lay a groundwork for a basic reconsideration of that composition dependence. For example, we don't insist here that $\mathcal{X}_{12}$ is the traditional Flory–Huggins χ_{12} parameter.

As with the composition dependence, the pressure dependence of χ_{12} has recently begun to receive more specific study (Beiner *et al.*, 1998). With regard

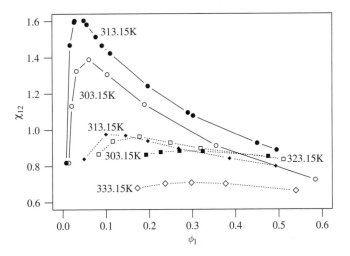

Figure 4.4 Composition dependence of the Flory–Huggins interaction parameter χ_{12} for physically distinct cases of mixtures of cyclohexane–polystyrene (dotted lines) and water–polyethylene glycol (solid lines), utilizing the data of (Bae *et al.*, 1993) for several temperatures. ϕ_1 is the volume fraction of the small-molecule solvent. For cyclohexane–polystyrene the assumption of composition independence of χ_{12} is qualitatively satisfactory; the lowest curve there is the $T = 60\,^\circ\text{C}$ data. For water–polyethylene glycol the assumption of composition independence of χ_{12} is less satisfactory; the upper curve there is the $T = 40\,^\circ\text{C}$ data, and the lower curve corresponds to $T = 30\,^\circ\text{C}$. The data were analyzed on the basis of Eq. (4.63), determining the value $\chi_{12}(\phi_1 = 0)$ by fitting the empirical data to the form $(\phi_1 - [\phi_1]_{\min})\chi_{12}(\phi_1 = 0)$, with $[\phi_1]_{\min}$ the minimum volume fraction measured, and regarding $\chi_{12}(\phi_1 = 0)$ as a fitting parameter. The temperature dependence is stronger, and thus clearer, for the water–polyethylene glycol case, and the interaction strength increases with increasing temperature, reflecting significant entropic contributions.

to temperature dependences, the original view was that $\chi_{12} \propto 1/T$. Though there are remarkable exceptions (Lefebvre *et al.*, 1999), it is an important experimental point that linearity in $1/T$ is common; a refinement is that linearity should be distinguished from proportionality, and this issue will come up in the discussion below.

Here we develop a derivation of the Flory–Huggins model that is broader than the customary derivations. The derivation below doesn't, at an initial stage, express various quantities related to polymeric materials on a per monomer basis, as is customary. The chief reason for this is that such a definition at an initial stage is typically based upon lattice modeling of the problem, and we wish here to avoid premature idealizations.

We consider the mixing of two liquids. We will assume, in the first place, that a satisfactory separation of attractive and repulsive intermolecular interactions is possible, and that the attractive interactions can be treated in the usual van der Waals approach of Section 4.1, p. 61. Thus, the interaction contribution to chemical potentials, or the *excess* chemical potentials, of each species involved will be assumed to adopt the form

$$\beta\mu_1^{ex}(\rho_1, \rho_2) \approx \beta\tilde{\mu}_1^{ex}(\rho_1, \rho_2) - \frac{2}{kT}\sum_2 a_{12}^{(2)}\rho_2. \qquad (4.40)$$

The $a_{12}^{(2)}$ are van der Waals parameters describing the effects of attractive interactions, and the tilde with $\tilde{\mu}_1^{ex}(\rho_1, \rho_2)$ indicates that these are the excess chemical potentials for the case *without* attractive interactions, i.e. the packing contributions associated with the repulsive forces solely. The experimental results replotted in Fig. 4.5 show that the $1/T$ dependence implied by this assumption can be satisfactory.

We first consider packing contributions and compose the Gibbs free energy of mixing for that contribution as

$$\frac{\beta\Delta\tilde{\mathcal{G}}_{mix}}{N} = x_1\left[\ln\frac{\rho_1}{\bar{\rho}_1} + \beta\tilde{\mu}_1^{ex}(\rho_1, \rho_2) - \beta\tilde{\mu}_1^{ex}(\bar{\rho}_1, 0)\right]$$
$$+ x_2\left[\ln\frac{\rho_2}{\bar{\rho}_2} + \beta\tilde{\mu}_2^{ex}(\rho_1, \rho_2) - \beta\tilde{\mu}_2^{ex}(0, \bar{\rho}_2)\right]. \qquad (4.41)$$

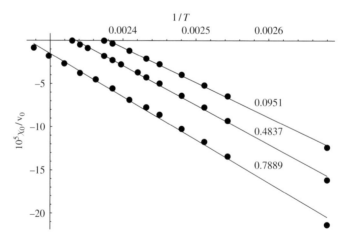

Figure 4.5 Measurements of *effective* Flory–Huggins χ parameter for deuterated polystyrene–poly(vinylmethylether) blends (Han *et al.*, 1988) as a function of $1/T$. The temperature range is typically 100–150 °C, and the numbers near each set of data indicate the volume fraction of polystyrene. The observed linear dependence on $1/T$ provides the physical conclusion that first-order perturbation theory is a satisfactory treatment of attractive interactions here.

The quantities $\bar{\rho}_1$ are the densities of the pure phases at the specified pressure p, that is, with attractive forces operating to achieve experimental densities; the partial molar volumes of these pure phases are $\bar{v}_1 = 1/\bar{\rho}_1$. Similarly, the densities of the mixture correspond to the same pressure p. Remembering Eq. (4.39), a way for this mixing free energy, Eq. (4.41), to achieve the form of the initial two terms of Eq. (4.38) is that the excess quantities within the brackets of Eq. (4.41) vanish:

$$\beta\tilde{\mu}_1^{\text{ex}}(\phi_1\bar{\rho}_1, \phi_2\bar{\rho}_2) = \beta\tilde{\mu}_1^{\text{ex}}(\bar{\rho}_1, 0), \tag{4.42a}$$

$$\beta\tilde{\mu}_2^{\text{ex}}(\phi_1\bar{\rho}_1, \phi_2\bar{\rho}_2) = \beta\tilde{\mu}_2^{\text{ex}}(0, \bar{\rho}_2). \tag{4.42b}$$

To see how this might happen, look at these equations through the level of second virial coefficients

$$\tilde{b}_{11}^{(2)}\phi_1\bar{\rho}_1 + \tilde{b}_{12}^{(2)}\phi_2\bar{\rho}_2 = \tilde{b}_{11}^{(2)}\bar{\rho}_1, \tag{4.43a}$$

$$\tilde{b}_{21}^{(2)}\phi_1\bar{\rho}_1 + \tilde{b}_{22}^{(2)}\phi_2\bar{\rho}_2 = \tilde{b}_{22}^{(2)}\bar{\rho}_2. \tag{4.43b}$$

The tilde is a reminder that we are discussing packing contributions here. Solving these equations produces

$$\tilde{b}_{11}^{(2)} = \frac{\bar{v}_1\phi_2}{\bar{v}_2(1-\phi_1)}\tilde{b}_{12}^{(2)}, \tag{4.44a}$$

$$\tilde{b}_{22}^{(2)} = \frac{\bar{v}_2\phi_1}{\bar{v}_1(1-\phi_2)}\tilde{b}_{12}^{(2)}. \tag{4.44b}$$

If the excess volume of mixing is zero, then all the volume fraction multipliers cancel out in these forms, because then $\phi_1 = (1-\phi_2)$. Though the present derivation should provide a basis for doing better, this assumption of zero volume of mixing will be our second assumption here. Then Eqs. (4.44) describe a scaling of second virial coefficients that is natural even though special:

$$\tilde{b}_{11}^{(2)} = \left(\frac{\bar{v}_1}{\bar{v}_2}\right)\tilde{b}_{12}^{(2)}, \tag{4.45a}$$

$$\tilde{b}_{22}^{(2)} = \left(\frac{\bar{v}_2}{\bar{v}_1}\right)\tilde{b}_{12}^{(2)}. \tag{4.45b}$$

If species 1 is a solvent non-macromolecule of roughly the same size as the monomers composing the polymeric species, then it is natural to estimate $\bar{v}_2/\bar{v}_1 \approx M$, polymerization index. Further, it is natural also to estimate $\tilde{b}_{12}^{(2)}/\tilde{b}_{11}^{(2)} \approx M$; see Eq. (4.34), p. 77.

We now turn to discuss the treatment of attractive interactions. Again, our first discussion will be confined to a second virial coefficient level. We will use the notation

$$b_{12}^{(2)}(T) = \tilde{b}_{12}^{(2)} - \frac{a_{12}^{(2)}}{kT}. \tag{4.46}$$

This is consistent with Eq. (4.40) provided the $a_{12}^{(2)}$ introduced first were indeed independent of thermodynamic state. To compose the mixing free energy, note that

$$x_\eta = \frac{\phi}{\phi_1 \bar{\rho}_1 + \phi_2 \bar{\rho}_2}. \tag{4.47}$$

We then find

$$\frac{\beta \Delta \mathcal{G}_{\text{mix}}}{N} = x_1 \ln \phi_1 + x_2 \ln(1 - \phi_1)$$

$$- \frac{2\phi_1(1 - \phi_1)}{kT} \left(\frac{\bar{\rho}_1^2 a_{11}^{(2)} - 2\bar{\rho}_2 \bar{\rho}_1 a_{12}^{(2)} + \bar{\rho}_2^2 a_{22}^{(2)}}{\phi_1(\bar{\rho}_1 - \bar{\rho}_2) + \bar{\rho}_2} \right) \tag{4.48}$$

provided Eq. (4.45), and assuming that the volume of mixing is zero. This Eq. (4.48) is recognizable as the form Eq. (4.38) if the interaction parameter is identified as

$$\mathcal{X}_{12} = -\frac{2}{kT} \left(\frac{\bar{\rho}_1^2 a_{11}^{(2)} - 2\bar{\rho}_2 \bar{\rho}_1 a_{12}^{(2)} + \bar{\rho}_2^2 a_{22}^{(2)}}{\phi_1(\bar{\rho}_1 - \bar{\rho}_2) + \bar{\rho}_2} \right). \tag{4.49}$$

If the various parameters $a_{12}^{(2)}$ were to scale as the repulsive force contributions have been assumed, Eq. (4.45), then this formula Eq. (4.49) would vanish. But the same scaling for attractive force parameters as for repulsive force parameters is not as reasonable. The attractive force contributions derive from longer-range interactions and relative strengths of those interactions may display additional variety. The calculation leading to Eq. (4.49) does, however, show that the slightly more general relation

$$\bar{\rho}_1^2 \tilde{b}_{11}^{(2)} - 2\bar{\rho}_2 \bar{\rho}_1 \tilde{b}_{12}^{(2)} + \bar{\rho}_2^2 \tilde{b}_{22}^{(2)} = 0 \tag{4.50}$$

is a sufficient restriction on the repulsive force contributions through the level of the second virial coefficients. Thus, through the level of second virial coefficients only, we could have written

$$\mathcal{X}_{12} = 2 \left(\frac{\bar{\rho}_1^2 b_{11}^{(2)} - 2\bar{\rho}_2 \bar{\rho}_1 b_{12}^{(2)} + \bar{\rho}_2^2 b_{22}^{(2)}}{\phi_1(\bar{\rho}_1 - \bar{\rho}_2) + \bar{\rho}_2} \right) \tag{4.51}$$

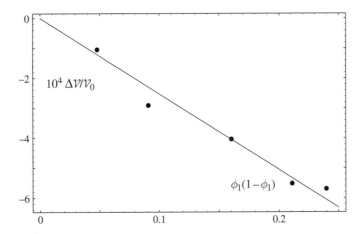

Figure 4.6 Volume of mixing of poly(ethymethylsiloxane) and poly(dime-thylsiloxane), a case for which this is particularly small. These results of Beiner *et al.* (1998) have been normalized by the specific volume of poly(dime-thylsiloxane) (Markovitz and Zapas, 1989, see Table 13.16.B.1.a) at $T = 30\,°C$. For other details see Beiner *et al.* (1998).

involving the full second virial coefficients, Eq. (4.46). This is a little less fruitful, however, because Eq. (4.45) is a valuable physical insight, and we should be able to exploit that further.

The preceding arguments have been specialized, so let's review them to see to what extent they might be generalized. Most importantly, we see that the conclusion from Eq. (4.43) can be much more general. The calculation of Eq. (4.43) assesses excluded volumes, and requires that the volume available to a distinguished additional molecule be the same for the pure solvent and the mixed solution. Theories of such packing effects can depend only on packing fractions evaluated by matching second virial coefficients for those effects. A simple example is the primordial free-volume theory $-\ln(1-\xi)$ where ξ is the packing fraction (Hildebrand, 1947); this primordial free volume $-\ln(1-\xi)$ is approximately equal to ξ for small values of that packing fraction, and that correspondence provides a fundamental identification of the packing fraction. Theories of this sort are widespread, typically among the more accurate theories available (Bjorling *et al.*, 1999), and the developments above apply to those theories. Indeed, the idea that distinguished particle packing should depend primarily on the net packing fraction, and only secondarily on structural idiosyncrasies, conforms more to common sense. Liquid crystalline structuring is probably the most immediate contrary case; comparing fluid to liquid crystalline systems at the same packing fraction, it is clear that the structural distinctions make a qualitative difference.

A further issue for review is the treatment of attractive interactions. The treatment here was limited to consideration of the second virial coefficient as in Eq. (4.46), and this implies the composition and temperature dependences exhibited in Eq. (4.49). Those composition and temperature dependences are certainly the leading factors, but a more general evaluation of first-order perturbation theory could result in subtle corrections to those dependences. Additionally, some implicit temperature and pressure dependence is implied by the variations of the pure liquid properties in those factors. Finally, the limitation of the first-order perturbation theory must also be borne in mind; there are experimental cases where first-order perturbation theory appears to be unsatisfactory (Lefebvre *et al.*, 1999).

The view here is broader than the classic view of the Flory–Huggins model. Though consistent with the qualitative message of traditional Flory–Huggins treatments, the requirement Eq. (4.45) will be used in developing subsequent corrections to the most primitive Flory–Huggins treatments. The required scaling of virial coefficients is reasonable for polymeric mixtures. The requirement on the volume of mixing is based upon the experience that the volume of mixing of polymeric liquids is typically small. Neither of these requirements need be the case for other liquid mixtures of molecules of different sizes, e.g. $(C_3F_7COOCH_2)_4C$ and I_2 (Hildebrand *et al.*, 1970). Together these restrictions amount to a common sense, but approximate procedure for avoiding dominating packing issues.

We can formalize the discussion above, perhaps at the loss of some molecular-scale insight, by making replacements of $b_{12}^{(2)}$ according to consideration of Eqs. (4.42) on the basis of

$$\tilde{\mu}_1^{ex}(\phi_1\bar{\rho}_1, \phi_2\bar{\rho}_2) \approx \tilde{\mu}_1^{ex}(\bar{\rho}_1, 0)$$

$$+ (\phi_1 - 1)\bar{\rho}_1 \frac{\partial \tilde{\mu}_1^{ex}}{\partial \rho_1}(\bar{\rho}_1, 0) + \phi_2\bar{\rho}_2 \frac{\partial \tilde{\mu}_1^{ex}}{\partial \rho_2}(\bar{\rho}_1, 0), \quad (4.52a)$$

$$\tilde{\mu}_2^{ex}(\phi_1\bar{\rho}_1, \phi_2\bar{\rho}_2) \approx \tilde{\mu}_2^{ex}(0, \bar{\rho}_2)$$

$$+ \phi_1\bar{\rho}_1 \frac{\partial \tilde{\mu}_2^{ex}}{\partial \rho_1}(0, \bar{\rho}_2) + (\phi_2 - 1)\bar{\rho}_2 \frac{\partial \tilde{\mu}_2^{ex}}{\partial \rho_2}(0, \bar{\rho}_2). \quad (4.52b)$$

From the point of view of molecular theory, the coefficients $\partial \mu_1^{ex}/\partial \rho_2$ are fundamentally related to structural properties of these fluids – OZ direct correlation functions (Eq. (6.71), p. 141) – as is discussed in detail subsequently in Section 6.3 on the Kirkwood–Buff theory. Alternatively, these coefficients could be explicitly evaluated if an explicit statistical thermodynamic model, as in the discussion here, were available for the unmixed fluids. Finally, these comments indicate that much of the information supplied by these coefficients is susceptible to measurement on and modeling of the unmixed liquids, as for example

$$\bar{\rho}_1 \frac{\partial \tilde{\mu}_1^{ex}}{\partial \rho_1}(\bar{\rho}_1, 0) = \frac{\partial \tilde{p}^{ex}}{\partial \rho_1} = \frac{\partial \tilde{p}}{\partial \rho_1} - kT$$

$$\approx \frac{\partial p}{\partial \rho_1} + 2a_{11}^{(2)}\rho_1 - kT \tag{4.53}$$

for the pure fluid. This gives further perspective on the appearance of pure-fluid compressibilities in formulae such as Eq. (4.57) below.

We now specialize these general formulae to the common case that this mixing free energy is expressed on a per monomer basis. As a notational convenience we consider specifically the case of a polymer species mixed into a small molecule solvent, and use $M = \bar{v}_2/\bar{v}_1$ as the empirical polymerization index parameter. Then

$$\frac{\beta \Delta \mathcal{G}_{mix}}{\bar{\rho}_1 V} = \phi_1 \ln \phi_1 + \frac{(1-\phi_1)}{M} \ln(1-\phi_1) \tag{4.54}$$

is the ideal contribution to the mixing free energy on a per monomer basis. The quantity $\bar{\rho}_1 V$ would be the nominal number of monomers that would occupy the volume V under the conditions considered. For the interaction contribution the required conversion is

$$\frac{N}{\bar{\rho}_1 V} = \phi_1 + \frac{1}{M}\phi_2, \tag{4.55}$$

which is directly proportional to the denominator of Eq. (4.68). Defining χ_{12} yields, then, the conventional form:

$$\frac{\beta \Delta \mathcal{G}_{mix}}{\bar{\rho}_1 V} = \phi_1 \ln \phi_1 + \frac{(1-\phi_1)}{M} \ln(1-\phi_1) + \chi_{12}\phi_1(1-\phi_1). \tag{4.56}$$

Notice that here the conventional Flory–Huggins χ_{12} appears rather than X_{12}. Figure 4.7 shows what this function of ϕ_1 looks like for several values of χ_{12}. Then if we accept the specific calculation of Eq. (4.68), for example,

$$-kT\bar{\rho}_1\chi_{12} = 2\bar{\rho}_1^2 \left(a_{11}^{(2)} - \frac{2a_{12}^{(2)}}{M} + \frac{a_{22}^{(2)}}{M^2} \right)$$

$$+ \left(\frac{1}{\bar{\kappa}_{T1}} - kT\bar{\rho}_1 + 2a_{11}^{(2)}\bar{\rho}_1^2 \right)\left(1 - \frac{\tilde{b}_{12}^{(2)}}{\tilde{b}_{11}^{(2)}} \frac{1}{M} \right)$$

$$+ \left(\frac{1}{\bar{\kappa}_{T2}} - kT\bar{\rho}_2 + 2a_{22}^{(2)}\bar{\rho}_2^2 \right)\left(1 - \frac{\tilde{b}_{12}^{(2)}}{\tilde{b}_{22}^{(2)}} M \right), \tag{4.57}$$

on a per monomer basis.

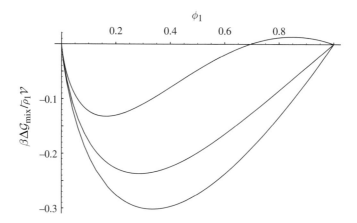

Figure 4.7 Examples of the function Eq. (4.56) for $M = 100$, and $\chi_{12} = 1/4, 1/2, 1$, from bottom to top.

Exercises

4.16 Consider a case of a mixture of two polymer species with nominal polymerization indices of M_γ and M_ν where the ideal free energy of mixing is to be described by

$$\frac{\phi_1}{M_1} \ln \phi_1 + \frac{\phi_2}{M_2} \ln \phi_2. \tag{4.58}$$

Suggest what the normalization for this mixing free energy might be.

4.17 Show that

$$\left[\frac{\partial \Delta \mathcal{G}_{\text{mix}}}{\partial n_1}\right]_{T,p,n_2} = \mu_1 - \bar{\mu}_1, \tag{4.59}$$

where the rightmost quantity is the indicated chemical potential for the pure liquid of species 1 at the prescribed temperature and pressure. Show further that

$$\left[\frac{\partial \phi_1}{\partial n_1}\right]_{T,p,n_2} = \frac{\phi_1 (1 - \phi_1)}{n_1}, \tag{4.60}$$

under the assumption that the volume of mixing vanishes. Then consider the Flory–Huggins expression for the ideal entropy of mixing, Eq. (4.54), and show that

$$\beta \mu_1 - \beta \bar{\mu}_1 = \ln \phi_1 + (1 - \phi_1)\left(1 - \frac{1}{M}\right). \tag{4.61}$$

Give an interpretation of the last term shown here.

4.18 Consider the conditions of equilibrium involving the chemical potentials for two phases at specified temperature and pressure. Show that a double tangent in the $\beta \Delta \mathcal{G}_{\text{mix}}/N$ vs. mole fraction (or volume fraction) curve indicates a phase separation in a binary liquid mixture. What is the physical origin for phase separation at the level of Flory–Huggins theory?

4.19 With natural assumptions of the Flory–Huggins theory, from Eq. (4.38) show that

$$\beta \mu_1 = \beta \bar{\mu}_1 + \ln \phi_1 + (1 - \phi_1) \left(1 - \frac{1}{M} \right)$$

$$+ (1 - \phi_1)^2 \chi_{12} + (1 - \phi_1)^2 \phi_1 \frac{\partial \chi_{12}}{\partial \phi_1}. \tag{4.62}$$

4.20 Equation (4.62) suggests an operational definition of an empirical χ parameter

$$\chi \equiv \frac{1}{(1 - \phi_1)^2} \left[\beta \mu_1 - \beta \bar{\mu}_1 - \ln \phi_1 - (1 - \phi_1) \left(1 - \frac{1}{M} \right) \right]$$

$$= \chi_{12} + \phi_1 \frac{\partial \chi_{12}}{\partial \phi_1}; \tag{4.63}$$

see Bae *et al.* (1993). Following Eq. (4.63), show that the Flory–Huggins χ_{12} can be obtained from the empirical χ according to the relation

$$\phi_1 \chi_{12} (\phi_1) = \int_0^{\phi_1} \chi(\phi) \mathrm{d}\phi. \tag{4.64}$$

4.21 Give a thermodynamic explanation as to why $b_{12}^{(2)} = b_{21}^{(2)}$. What can you say about this symmetry in the context of the formalization Eq. (4.52)?

4.22 Let's consider a specific development of the argument above. Suppose that each $\beta \tilde{\mu}_1(\rho_1, \rho_2)$ does have the form of a free-volume theory, thus being given by $\beta \tilde{\mu}_1(\rho_1, \rho_2) = m(\xi_1)$ where ξ_1 and ξ_2 are the packing fractions

$$\frac{1}{2} \xi_1 = \tilde{b}_{11}^{(2)} \phi_1 \bar{\rho}_1 + \tilde{b}_{12}^{(2)} \phi_2 \bar{\rho}_2, \tag{4.65a}$$

$$\frac{1}{2} \xi_2 = \tilde{b}_{21}^{(2)} \phi_1 \bar{\rho}_1 + \tilde{b}_{22}^{(2)} \phi_2 \bar{\rho}_2. \tag{4.65b}$$

and $m(x)$ is a given function such as $- \ln (1 - x)$. Suppose further that the virial coefficients don't quite obey Eq. (4.44) and consider the discrepancies

$$\frac{1}{2} \delta \xi_1 = -\tilde{b}_{11}^{(2)} (1 - \phi_1) \bar{\rho}_1 + \tilde{b}_{12}^{(2)} \phi_2 \bar{\rho}_2, \tag{4.66a}$$

$$\frac{1}{2} \delta \xi_2 = -\tilde{b}_{22}^{(2)} (1 - \phi_2) \bar{\rho}_2 + \tilde{b}_{12}^{(2)} \phi_1 \bar{\rho}_1. \tag{4.66b}$$

Assuming that the volume of mixing is zero, and that these discrepancies aren't large, write

$$m\left(2\tilde{b}_{11}^{(2)}\bar{\rho}_1 + \delta\xi_1\right) - m\left(2\tilde{b}_{11}^{(2)}\bar{\rho}_1\right) \approx m'\left(2\tilde{b}_{11}^{(2)}\bar{\rho}_1\right)\delta\xi_1$$

$$= -\phi_2\left(\frac{\tilde{b}_{11}^{(2)}\bar{\rho}_1 - \tilde{b}_{12}^{(2)}\bar{\rho}_2}{\tilde{b}_{11}^{(2)}\bar{\rho}_1}\right)\left(\frac{\beta}{\bar{\rho}_1\bar{\kappa}_T(1)} - 1 + 2\beta a_{11}^{(2)}\bar{\rho}_1\right). \quad (4.67)$$

The correction is the first term in a Taylor series for small discrepancies, and involves $m'(\xi_1)$, a property of the pure liquid that has been evaluated here in terms of the pure liquid compressibility, $\bar{\kappa}_{T1}$. The term $2\beta a_{11}^{(2)}\bar{\rho}_1$ appears because $m(x)$ describes only the packing contributions, and in exploiting the measured compressibility of the actual system we should correct for the influence of attractive interactions. We do that on the basis of a van der Waals model with parameter $a_{11}^{(2)}$ for pure fluid 1, compensating for a trivial temperature-dependent contribution in the $\bar{\kappa}_{T1}$. Thus the temperature dependences of those coefficients are intended to be weak. Including this correction, show that the revised strength of the interaction parameter is

$$-\mathcal{X}_{12} = \frac{2}{kT}\left(\frac{\bar{\rho}_1{}^2 a_{11}^{(2)} - 2\bar{\rho}_2\bar{\rho}_1 a_{12}^{(2)} + \bar{\rho}_2{}^2 a_{22}^{(2)}}{\phi_1(\bar{\rho}_1 - \bar{\rho}_2) + \bar{\rho}_2}\right)$$

$$+ \left[\left(\frac{\beta}{\bar{\rho}_1\bar{\kappa}_{T1}} - 1 + 2\beta a_{11}^{(2)}\bar{\rho}_1\right)\left(\bar{\rho}_1 - \frac{\tilde{b}_{12}^{(2)}}{\tilde{b}_{11}^{(2)}}\bar{\rho}_2\right)\right.$$

$$+ \left.\left(\frac{\beta}{\bar{\rho}_2\bar{\kappa}_{T2}} - 1 + 2\beta a_{22}^{(2)}\bar{\rho}_2\right)\left(\bar{\rho}_2 - \frac{\tilde{b}_{12}^{(2)}}{\tilde{b}_{22}^{(2)}}\bar{\rho}_1\right)\right]$$

$$/\left[\phi_1(\bar{\rho}_1 - \bar{\rho}_2) + \bar{\rho}_2\right]. \quad (4.68)$$

This is no longer proportional to $1/T$ but is linear in $1/T$, aside from dependences on pure fluid properties. The last term of Eq. (4.68) would be regarded as an entropic contribution to $\mathcal{X}_{12}$.

4.23 Show that the corrected result Eq. (4.68) matches Eq. (4.49) through the second virial coefficient contributions.

4.24 The discussion of this section introduced volume fractions

$$\phi_1 \equiv \frac{\bar{\rho}_1}{\tilde{\rho}_1}, \quad (4.69)$$

and at opportune steps assumed that the volume of mixing vanished, so $\phi_1 + \phi_2 = 1$. Reconsider these derivations permitting the volume of mixing to be non-zero,

$$\phi_1 + \phi_2 = 1 - \frac{\Delta \mathcal{V}}{\mathcal{V}},\tag{4.70}$$

and see what you can learn.

4.5 Electrolyte solutions and the Debye–Hückel theory

The distinction between drinking water and seawater is obvious. Electrolyte solutions, as the name suggests, involve neutral components that separate into electrically non-neutral units, *ions*. Ionic interactions are $q_\eta q_\nu / r$ for a pair of ions with formal charges q_η and q_ν separated by a distance r.

Typical ionic interactions are strong on a thermal energy scale. We can characterize the strength of these interactions by considering the distance between monovalent ions when their mutual interaction potential energy corresponds to the standard thermal energy $(T = 298\,\mathrm{K}) : r = \beta e^2 \approx 561\,\text{Å}$. Because these interactions are strong, electrolyte solutions are common for cases where a substantial solvent dielectric constant can weaken the net size of inter-ionic influences. For the common example of liquid water solvent, $\epsilon \approx 80$. Then it is common initially to consider models in which the effects of the solvent are subsumed by this dielectric screening (Friedman and Dale, 1977), and the inter-ionic interactions are described by $q_\eta q_\nu / \epsilon r$. Still, ion pairs closer than about $7\,\text{Å}$ have mutual interactions stronger then kT. Typical inter-ionic distances for solutes as concentrated as $1\,\mathrm{mol\,l}^{-1}$ of a 1-1 salt, such as NaCl, are about $9\,\text{Å}$, estimated as $\rho^{-1/3}$ where ρ is the number density of ions. To put these strength-of-interaction issues in the background, we restrict attention to significantly lower concentrations than this.

But even at low concentration difficulties remain: the characteristic difficulty of these interactions is their long range. One important consequence of the long range of these interactions is that the bulk compositions of electrolyte solutions are neutral. These solutions conduct electricity and the macroscopic electric field in the interior of a conductor at equilibrium must be zero. Otherwise, currents would flow, heat would be generated, and the system wouldn't be at thermal equilibrium. But if the macroscopic electric field is zero in an interior of a conductor at equilibrium, then the Poisson equation can be read in reverse as implying that the equilibrium macroscopic charge density must be zero. Viewed physically, a slight excess charge will be conducted until it reaches a surface without possibilities for further adjustment. This influence of the system surface is a clue that long-ranged interactions are involved.

These considerations can make the thermodynamics of electrolyte solutions tricky. Despite such possibilities, the right side of Eq. (3.18), p. 40, is typically inoffensive for the problems of ionic contributions to single-ion activities. In fact, early instances (Kirkwood and Poirier, 1954; Jackson and Klein, 1964) of the PDT were focused precisely on our present problem. The averaging need only involve a neutral solution and a decoupled distinguished ion. The requirements are that the potential energy change be well defined and that the sampling can be performed.

We will begin a more detailed discussion from the point of view of a gaussian model Eq. (4.12), p. 66. Because we will be interested in the lowest concentrations, we will neglect consideration of short-ranged interactions; they are necessary to define our problem but wouldn't appear in our final result here. Also because we are interested in the lowest concentrations (so the typical ionic interactions aren't too strong) and temperatures not too low (β not too high), the fact that the terms of Eq. (4.12), p. 66, exhibit a formal ordering in powers of β is also motivational. Thus,

$$\mu_\gamma^{\mathrm{ex}} \approx \tilde{\mu}_\gamma^{\mathrm{ex}} + \int_V \varphi_{\gamma\eta}(r) \langle \rho_\eta(r) \rangle_{\mathrm{r}} \, \mathrm{d}^3 r$$

$$- \frac{\beta}{2} \int_V \int_V \varphi_{\gamma\eta}(r) \langle \delta\rho_\eta(r)\delta\rho_\nu(r') \rangle_{\mathrm{r}} \varphi_{\nu\gamma}(r') \mathrm{d}^3 r \mathrm{d}^3 r' \qquad (4.71)$$

where we have dropped the notation of the condition $\mathcal{R}^n$ since we won't exploit information of that type in the present discussion. Here $\varphi_{\eta\nu}(r) = q_\eta q_\nu / \epsilon r$; $\langle \rho_\eta(r) \rangle_{\mathrm{r}}$ is the density of species η obtained for the reference system with a distinguished particle which interacts with the solution on the basis of the defined reference interactions.

The discussion surrounding Eq. (4.13), p. 66, suggested it was reasonable to drop the subscript qualifier on $\langle \ldots \rangle_{\mathrm{r}}$ in treating the longest-ranged interactions. Noting that in that case the middle term of Eq. (4.71) vanishes by electroneutrality of the bulk compositions, we then consider the simplification

$$\mu_\gamma^{\mathrm{ex}} \approx \tilde{\mu}_\gamma^{\mathrm{ex}} - \frac{\beta}{2} \int_V \int_V \varphi_{\gamma\eta}(r) \langle \delta\rho_\eta(r)\delta\rho_\nu(r') \rangle \varphi_{\nu\gamma}(r') \mathrm{d}^3 r \mathrm{d}^3 r'. \qquad (4.72)$$

This looks like the most primitive theory that we might justifiably take seriously. It depends on information about correlations

$$\langle \delta\rho_\eta(r)\delta\rho_\nu(r') \rangle = \rho_\nu \delta_{\nu\eta}\delta(r-r') + \rho_\nu\rho_\eta \left(g_{\nu\eta}^{(2)}(|r-r'|) - 1 \right). \qquad (4.73)$$

This type of information might be obtained as data from simulation or some other experiment.

But the PDT permits the same tools to be applied also to study $g_{\nu\eta}^{(2)}(|r-r'|)$; see Eq. (3.50), p. 49, which we rewrite for the specific case here:

$$g_{\nu\eta}^{(2)}(r,r') = \left(\frac{z_\nu}{\rho_\nu \Lambda_\nu^3}\right)\left(\frac{z_\eta}{\rho_\eta \Lambda_\eta^3}\right)e^{-\beta u_{\nu\eta}^{(2)}(r,r')}$$

$$\times \left\langle e^{-\beta\Delta U_{\nu\eta}^{(2)}}|r_{1\nu}=r, r_{1\eta}=r'\right\rangle_0. \tag{4.74}$$

The thermodynamic multipliers in Eq. (4.74) play the role of achieving $g_{\nu\eta}^{(2)}(|r-r'|) \sim 1$ for large $|r-r'|$; see Eq (3.50) p. 49.

For simplicity in the present context, we will assume that the microscopic binding energies of the two distinguished ions are additive according to $\Delta U_{\nu\eta}^{(2)} = \Delta U_\nu^{(1)} + \Delta U_\eta^{(1)}$. If the variables $\Delta U_\nu^{(1)}$ and $\Delta U_\eta^{(1)}$ were uncorrelated, then because of the normalization achieved by the thermodynamic multipliers, Eq. (4.74) would yield $g_{\nu\eta}^{(2)}(r,r') = e^{-\beta u_{\nu\eta}^{(2)}(r,r')}$. This is not satisfactory in the present case because of the long range of the inter-ionic interactions $u_{\nu\eta}^{(2)}(r,r')$. Therefore, the effects that we are after here involve the correlations of these variables.

We can make this issue of correlations more definite by writing

$$\left\langle e^{-\beta(\Delta U_\nu^{(1)}+\Delta U_\eta^{(1)})}|r_{1\nu}=r, r_{1\eta}=r'\right\rangle_0$$

$$= \int \mathcal{P}_{\nu\eta}^{(0)}\left(\varepsilon, \varepsilon'|r_{1\nu}=r, r_{1\eta}=r'\right)e^{-\beta(\varepsilon+\varepsilon')}d\varepsilon d\varepsilon' \tag{4.75}$$

where

$$\mathcal{P}_{\nu\eta}^{(0)}\left(\varepsilon, \varepsilon'|r_{1\nu}=r, r_{1\eta}=r'\right)$$

$$= \left\langle \delta\left(\Delta U_\nu^{(1)}-\varepsilon\right)\delta\left(\Delta U_\eta^{(1)}-\varepsilon'\right)|r_{1\nu}=r, r_{1\eta}=r'\right\rangle_0. \tag{4.76}$$

Introducing the marginal distributions

$$\mathcal{P}_\nu^{(0)}(\varepsilon) = \int \mathcal{P}_{\nu\eta}^{(0)}\left(\varepsilon, \varepsilon'|r_{1\nu}=r, r_{1\eta}=r'\right)d\varepsilon', \tag{4.77}$$

then the uncorrelated approximation is

$$\mathcal{P}_{\nu\eta}^{(0)}\left(\varepsilon, \varepsilon'|r_{1\nu}=r, r_{1\eta}=r'\right) \approx \mathcal{P}_\nu^{(0)}(\varepsilon)\mathcal{P}_\eta^{(0)}(\varepsilon'), \tag{4.78}$$

without dependence on the separation $|r-r'|$.

Now the statistical model that underlies Eq. (4.72) is a gaussian one. Thus we anticipate that a gaussian model should be consistent for $\mathcal{P}_{\nu\eta}^{(0)}\left(\varepsilon, \varepsilon'|r_{1\nu}=r, r_{1\eta}=r'\right)$ although we should exclude uncorrelated features of the joint gaussian distribution. On the basis of Eq. (4.78) those uncorrelated features just establish an

$|r - r'|$ independent normalization. Performing that calculation (see Eq. (4.15), p. 67) gives

$$-\beta^{-1} \ln g^{(2)}_{\nu\eta}(r) = u^{(2)}_{\nu\eta}(r) - \beta \left\langle \delta\Delta U^{(1)}_{\nu} \delta\Delta U^{(1)}_{\eta} | r_{1\nu} = 0, r_{1\eta} = r \right\rangle_0. \quad (4.79)$$

The $\delta\Delta U^{(1)}_{\nu}$ here means the usual thing:

$$\delta\Delta U^{(1)}_{\nu} = \Delta U^{(1)}_{\nu} - \left\langle \Delta U^{(1)}_{\nu} | r_{1\nu} = 0, r_{1\eta} = r \right\rangle_0. \quad (4.80)$$

In contrast with formulae like Eq. (4.72), p. 90, there is no surviving factor of 1/2 with the fluctuation term of Eq. (4.79). This fluctuation contribution comes from a quadratic cross-term and the customary factor of 2 with that cross-term has cancelled the 1/2; again see Eq. (4.15), p. 67.

In approaching a final result here, we emphasize again that this model must be expected to be unsatisfactory for the shortest-range correlations. For example, it cannot be expected to be satisfactory if $\Delta U^{(1)}_{\nu}$ describes van der Waals excluded volume interactions. With this restriction in mind, and with our physical orientation on this problem, we replace $\Delta U^{(1)}_{\nu} = \Phi^{(1)}_{\nu}$, so that Eqs. (4.79) and (4.80) are replaced by

$$-\beta^{-1} \ln g^{(2)}_{\nu\eta}(r) = u^{(2)}_{\nu\eta}(r) - \beta \left\langle \delta\Phi^{(1)}_{\nu} \delta\Phi^{(1)}_{\eta} | r_{1\nu} = 0, r_{1\eta} = r \right\rangle_0. \quad (4.81)$$

and

$$\delta\Phi^{(1)}_{\nu} = \Phi^{(1)}_{\nu} - \left\langle \Phi^{(1)}_{\nu} | r_{1\nu} = 0, r_{1\eta} = r \right\rangle_0. \quad (4.82)$$

Translating this result into the notation of Eq. (4.72), we have

$$\ln g^{(2)}_{\nu\eta}(r) \approx -\beta\varphi_{\nu\eta}(r)$$
$$+ \beta^2 \iint_{V V} \varphi_{\gamma\eta}(r') \left\langle \delta\rho_{\eta}(r') \delta\rho_{\nu}(r'') \right\rangle \varphi_{\nu\gamma}(r'' - r) d^3 r' d^3 r''. \quad (4.83)$$

We emphasize again that this is only reasonable at long range where $g^{(2)}_{\nu\eta}(r)$ tends to one (1) and $h^{(2)}_{\nu\eta}(r) = g^{(2)}_{\nu\eta}(r) - 1$ tends to zero (0). It is then natural to linearize according to $\ln\left[1 + h^{(2)}_{\nu\eta}(r)\right] \approx h^{(2)}_{\nu\eta}(r)$ which yields the classic Debye–Hückel theory:

$$h^{(2)}_{\nu\eta}(r) \approx -\beta\varphi_{\nu\eta}(r)$$
$$+ \beta^2 \iint_{V V} \varphi_{\gamma\eta}(r') \left\langle \delta\rho_{\eta}(r') \delta\rho_{\nu}(r'') \right\rangle \varphi_{\nu\gamma}(r'' - r) d^3 r' d^3 r''. \quad (4.84)$$

The quantity $\left\langle \delta\rho_{\eta}(r') \delta\rho_{\nu}(r'') \right\rangle$ on the right side involves $h^{(2)}_{\nu\eta}(r)$, so this is an integral equation to be solved for $h^{(2)}_{\nu\eta}(r)$.

The linearization that leads here to the Debye–Hückel model is physically consistent in this argument. But the possibility of a model that is unlinearized in this sense is a popular query. More than one response has been offered including the (nonlinear) Poisson–Boltzmann theory and the EXP approximation; see (Stell, 1977) also for representative numerical results for the systems discussed here.

In concluding this section, we note that the physical arguments here are involved for a subject (the Debye–Hückel theory) that is at once so basic, so firmly established, and so limited in physical scope to molecular science. The traditional presentation (e.g. Hill, 1986; Lewis *et al.*, 1961, see Section 23) is fine as far as it goes but gives little support for extensions of the theory, and little perspective on the basic issues of the theory of solutions. The physical discussion here is different from the most conventional presentations, does give further perspective on the role of the PDT, but is too extended without other pieces of the theory of solutions in place. All these points surely mean that this is one area where the beautiful but more esoteric theoretical tools (Lebowitz *et al.*, 1965) of professional theory of liquids are relevant to a simple view of the problem. This topic is taken up again after the developments of Section 6.1; see Eq. (6.28), p. 132.

Exercises

4.25 Take the model

$$h_{\alpha\gamma}(r) \approx -\frac{\beta q_\alpha q_\gamma e^{-\kappa r}}{\epsilon r} \qquad (4.85)$$

where

$$\kappa^2 \equiv \frac{4\pi\beta}{\epsilon} \sum_\gamma \rho_\gamma q_\gamma{}^2 \qquad (4.86)$$

and show that the fluctuation contribution to Eq. (4.72) is $-\left(\frac{q_\alpha{}^2 \kappa}{2\epsilon}\right)$ (Hill, 1986). We have emphasized that fluctuation contributions, e.g. Eq. (4.71) p. 90, have a definite sign. This Debye–Hückel theory treats correlations between ionic species, and here we observe again that treatment of correlations lowers this free energy.

4.26 Consider the converse of the question addressed in the previous exercise, i.e. in what sense is the identification of κ (Eq. (4.86)) necessary? Assume that Eq. (4.72) is satisfactory and that the correlations can be described with a simple screening length as in Eq. (4.85). Discuss why κ must be given by Eq. (4.86).

4.27 The simplest inclusion of a reference system contribution would be

$$\mu_\gamma^{\text{ex}} \approx 2kT \sum_\nu \tilde{b}_{\gamma\nu}\rho_\nu - \frac{q_\gamma{}^2 \kappa}{2\epsilon}, \qquad (4.87)$$

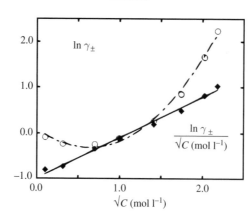

Figure 4.8 Chemical potential data for a primitive model 1-1 electrolyte from Valleau and Cohen (1980). See Eq. (4.87). The upper results are $\ln \gamma_\pm$, and the dashed-dot curve is a parabola fitted to those results. The lower results are $\ln \gamma_\pm / \sqrt{C}$ and the solid line is $-0.99 + 0.89\sqrt{C}$ fitted to those results; C is in mol l^{-1}.

where $\tilde{b}_{\gamma\nu}$ are expected to be positive, reflecting dominance of repulsive inter-ionic interactions for the reference system. Consider the data shown in Fig. 4.8 in the light of this simplest model. The fitted straight line gives $\ln \gamma_\pm = 0.89C - 0.99\sqrt{C}$, where C is the molar concentration of the salt. Evaluate the coefficients in Eq. (4.87), and check to what extent that model provides a satisfactory description of these data.

4.6 Clustering in dilute solutions and Pitzer models

Here we discuss a thermodynamic model appropriate to describe effects of strong association in dilute solutions. To have a definite example, consider a dilute electrolyte solution of a salt, say M_aX_b, that in solution dissociates to produce cations M of charge $q_M|e|$ and anions X of charge $-q_X|e|$ with $aq_M = bq_X$. The interactions between these ions are composed of short-ranged interactions and long-ranged ionic interactions screened by the dielectric response of the solvent with dielectric constant ϵ, as $q_M q_X / \epsilon r$, with r the distance between the ions. If the ionic charges are high or if the solvent dielectric constant isn't particularly high, then such interactions can lead to ion pairing, or clustering, and it is to models of this association phenomenon that we direct our attention. The model discussed here is a simplification (Pratt and LaViolette, 1998) of a quasi-chemical theory developed later but is similar to an informal physical treatment of ionic equilibria in steam given by Pitzer (1982; 1983).

Clustering: partitioning statistical possibilities

To seek ion clusters that might form, we need an operational definition useful in identifying them. Let's direct our attention to a specific, *distinguished* M ion. We will presume that a rule is available to determine how many other ions, probably X ions, are clustered with the distinguished M ion (Lewis *et al.*, 1961, see discussion "What do we mean by degree of dissociation?," pp. 307–308). If there are k of those counter-ions clustered, we would consider this M to be the nucleus of an MX_k cluster.

Similarly, all other M ions could be considered in turn and we could determine the number, k, of counter-ions each M ion possesses in its cluster. In this way we could determine the number of M ions having precisely k counter-ion partners. We will denote the fraction of M ions with k counter-ion partners by x_k. (It isn't important to our formal development whether some of those X ions might be involved in more than one MX_k cluster.) If we persist in focusing on a sole distinguished M ion, x_k is expected to be the fraction of time in a long record during which the ion has k clustered X partners. The size of these clusters will be limited by the maximum coordination number of the distinguished central particle. We will use the language that an MX_k cluster is a *cluster of size k ligands*.

This clustering idea seeks to partition our difficulties according to the numbers of ligands, k, that might be associated with a distinguished M ion. The basic equation

$$\langle n_{\text{M}} \rangle = \sum_k \langle n_{\text{MX}_k} \rangle = \frac{\mathcal{V} q_{\text{M}}^{\text{int}} z_{\text{M}}}{\Lambda_{\text{M}}^3} \left\langle\!\!\left\langle e^{-\beta \Delta U_{\text{M}}^{(1)}} \right\rangle\!\!\right\rangle_0 \tag{4.88}$$

will be a target of our present discussion; see Eq. (3.18) p. 40. The sum is over all MX_k clusters that can form on the distinguished M nucleus. This partition being established, it offers a *divide-and-conquer* possibility; calculations/measurements can be focused on a small number of cases, and then the net result composed according to Eq. (4.88).

To make progress with a specific term, say $\langle n_{\text{MX}_k} \rangle$, we assume that the PDT applies to this species too:

$$\langle n_{\text{MX}_k} \rangle = \frac{\mathcal{V} q_{\text{MX}_k}^{\text{int}} z_{\text{MX}_k}}{\Lambda_{\text{MX}_k}^3} \left\langle\!\!\left\langle e^{-\beta \Delta U_{\text{MX}_k}^{(1)}} \right\rangle\!\!\right\rangle_0. \tag{4.89}$$

This argument is examined and supported in more detail later, in Section 7.2 p. 146. The quantities $q_{\text{MX}_k}^{\text{int}}$ and $\Lambda_{\text{MX}_k}^3$ are proposed to be factors of the canonical partition function of an MX_k molecule or cluster, as in Eq. (3.16) p. 39. Our further assignment

$$z_{MX_k} = z_M z_X{}^k \tag{4.90}$$

is perhaps less obvious, and it, too, will be scrutinized in a subsequent chapter. Here we note that it is an assertion of chemical equilibrium

$$\mu_{MX_k} = \mu_M + k\mu_X \tag{4.91}$$

for the transformation

$$M + kX \rightleftharpoons MX_k. \tag{4.92}$$

Altogether we then obtain

$$\left\langle\!\!\left\langle e^{-\beta \Delta U_M^{(1)}} \right\rangle\!\!\right\rangle_0 = \sum_k \left(\frac{q_{MX_k}^{\text{int}} \Lambda_M^3}{\Lambda_{MX_k}^3 q_M^{\text{int}}} \right) \left\langle\!\!\left\langle e^{-\beta \Delta U_{MX_k}^{(1)}} \right\rangle\!\!\right\rangle_0 z_X^k. \tag{4.93}$$

This should be compared with Eq. (7.19), p. 149. The primitive quasi-chemical approximation is obtained from Eq. (4.93) upon setting

$$\left\langle\!\!\left\langle e^{-\beta \Delta U_{MX_k}} \right\rangle\!\!\right\rangle_0 \approx \left[\left\langle\!\!\left\langle e^{-\beta \Delta U_X} \right\rangle\!\!\right\rangle_0 \right]^k. \tag{4.94}$$

This expresses the physical view that the M ion is well-buried and thus not affected by the material exterior to the cluster, and that the conditions of the ligands are about the same whether bound in the cluster or nonspecifically considered. The summand of Eq. (4.93) then simplifies to

$$\left\langle\!\!\left\langle e^{-\beta \Delta U_{MX_k}} \right\rangle\!\!\right\rangle_0 z_X^k \approx \left(\frac{\rho_X \Lambda_X^3}{q_X^{\text{int}}} \right)^k. \tag{4.95}$$

Finally we form the standard combination of the left side of Eq. (3.18), p. 40:

$$\frac{\rho_M \Lambda_M^3}{z_M q_M^{\text{int}}} \approx 1 + \sum_{k \geq 1} K_k^{(0)}(T) \rho_X^k, \tag{4.96}$$

where

$$K_k^{(0)}(T) = \frac{q_{MX_k}^{\text{int}}/\Lambda_{MX_k}^3}{\left(q_M^{\text{int}}/\Lambda_M^3\right)\left(q_X^{\text{int}}/\Lambda_X^3\right)^k}. \tag{4.97}$$

These coefficients are just the equilibrium ratios (Eq. (2.8), p. 25) for chemical conversions Eq. (4.92) evaluated ideally. This primitive quasi-chemical model

$$\beta \mu_{\mathrm{M}}^{\mathrm{ex}} \approx -\ln \left[1 + \sum_{k \geq 1} K_k^{(0)}(T) \rho_{\mathrm{X}}^k \right], \tag{4.98}$$

is a simple sum over compositional possibilities for binding to an M ion, and thus has the feel of a reduced partition function. Being based upon the approximation Eq. (4.94), it neglects many of the physical contributions that have been discussed previously in this chapter; those preceding results could be marshaled to improve the approximation Eq. (4.94). Also, our discussion here hasn't worried about the possibilities that the M ion might interact with things other than the X ions that bind to it. These further issues are necessary in realistic applications. A subsequent chapter will establish quasi-chemical results more fully so that these important issues can be addressed in a more organized way.

Exercises

Pressure variation of solvation free energies

4.28 The variation of the solvation free energies with pressure is the partial molar volume and gives direct information on hydration structure. Consider a solute species such as the ion M above, diluted in a solvent denoted by W, for example, water. Recalling the chemical potential expression of Eq. (3.3), p. 33, show that the partial molar volume is

$$v_{\mathrm{M}} \equiv \left(\frac{\partial \mu_{\mathrm{M}}}{\partial p} \right)_{\beta,n}$$

$$= \frac{1}{\rho_{\mathrm{M}}} \left(\frac{\partial \rho_{\mathrm{M}}}{\partial \beta p} \right)_{\beta,n} + \left(\frac{\partial \beta \mu_{\mathrm{M}}^{\mathrm{ex}}}{\partial \rho_{\mathrm{W}}} \right)_{\beta,n} \left(\frac{\partial \rho_{\mathrm{W}}}{\partial \beta p} \right)_{\beta,n}. \tag{4.99}$$

Then confine interest to conditions of infinite dilution and show that

$$\lim_{\rho_{\mathrm{M}} \to 0} v_{\mathrm{M}} = (\kappa_T/\beta) \left[1 + \rho_{\mathrm{W}} \left(\frac{\partial \beta \mu_{\mathrm{M}}^{\mathrm{ex}}}{\partial \rho_{\mathrm{W}}} \right)_\beta \right], \tag{4.100}$$

where $\kappa_T = (-1/\mathcal{V})(\partial \mathcal{V}/\partial p)_T$ is the isothermal coefficient of bulk compressibility of the pure solvent.

4.29 Consider the model Eq. (4.98) where W molecules may complex an M solute according to

$$\mathrm{M} + m\mathrm{W} \rightleftharpoons \mathrm{MW}_m. \tag{4.101}$$

Show that the interaction contribution to v_M is

$$\lim_{\rho_M \to 0} v_M^{ex} = -\left(\frac{\rho_W \kappa_T}{\beta}\right)\left(\frac{\partial}{\partial \rho_W}\right)_\beta \ln\left[1 + \sum_{m \geq 1} K_m^{(0)} \rho_W^m\right]$$

$$= -\left(\frac{\rho_W \kappa_T}{\beta}\right) v_W \bar{m}, \qquad (4.102)$$

where $v_W = 1/\rho_W$ is the partial molar volume of the pure solvent (water) and

$$\bar{m} = \left(\frac{\partial}{\partial \ln \rho_W}\right)_\beta \ln\left[1 + \sum_{m \geq 1} K_m^{(0)} \rho_W^m\right] = \sum_{m \geq 0} m x_m. \qquad (4.103)$$

Temperature variation of solvation free energies

4.30 The temperature variation of the solvation free energy is the partial molar entropy and, because of its interpretation as an indicator of disorder, is of wide interest. As above, we focus here on the conditions of infinite dilution of a solute M in a W solution. Show that the interaction contribution to the partial molar entropy is

$$\lim_{\rho_M \to 0} s_M^{ex} = \left[\left(\frac{\partial \rho_W}{\partial T}\right)_p \left(\frac{\partial}{\partial \rho_W}\right)_T + \left(\frac{\partial}{\partial T}\right)_{\rho_W}\right]$$

$$\times \left(\frac{1}{\beta} \ln\left[1 + \sum_{m \geq 1} K_m^{(0)} \rho_W^m\right]\right). \qquad (4.104)$$

The first term on the right side accounts for the temperature dependence of the solvent density; that brings in the coefficient of thermal expansion for the pure solvent $(1/\mathcal{V})(\partial \mathcal{V}/\partial T)_p = \alpha_p$, and then requires the density derivative of the quasi-chemical contributions. But that density derivative was analyzed above when we considered the partial molar volume. Using those results, show that

$$\lim_{\rho_M \to 0} s_M^{ex}/k = -\bar{m}\left(T\alpha_p\right) + \left(\frac{\partial}{\partial T}\right)_{\rho_W}\left(T \ln\left[1 + \sum_{m \geq 1} K_m^{(0)} \rho_W^m\right]\right)$$

$$= -\bar{m}\left(T\alpha_p\right) + \left(1 - \beta\left(\frac{\partial}{\partial \beta}\right)_{\rho_W}\right)\ln\left[1 + \sum_{m \geq 1} K_m^{(0)} \rho_W^m\right]. (4.105)$$

Eq. (4.105) highlights chemical contributions to this entropy because

$$\left(\frac{\partial}{\partial \beta}\right)_{\rho_W} \ln\left[1 + \sum_{m \geq 1} K_m^{(0)} \rho_W^m\right] = -\sum_{m \geq 0} x_m \Delta H_m^{(0)}. \qquad (4.106)$$

Obtain this equation and explain the heats that appear. Eq. (4.105) then becomes

$$\lim_{\rho_M \to 0} s_M^{ex}/k = -\bar{m} \left(T\alpha_p \right) + \ln \left[1 + \sum_{m \geq 1} K_m^{(0)} \rho_W^m \right]$$

$$+ \beta \sum_{m \geq 0} x_m \Delta H_m^{(0)}. \tag{4.107}$$

5

Generalities

The following is a survey of some general and useful relations for evaluating chemical potentials and free energy changes. The number of such relations isn't large, but an overview is warranted here. Evaluations of free energy changes are typically the most basic and convincing validations of molecular simulation research. Calculations of free energy changes are typically more specialized undertakings than unspecialized simulations. If the problem at hand has been well-considered and calculations are to be specially directed to evaluate free energy changes, then thermodynamic or coupling parameter integration procedures are likely to be the most efficient possibilities. They are favorably stratified, they can have low bias, it is clear how computational effort can be added effectively as results accumulate, and they can be embarrassingly parallel. Other methods considered here, such as importance sampling and overlap methods, can be incorporated into thermodynamic integration methods, and can improve the results.

Nevertheless, there are cases where the alternatives to thermodynamic integration would be chosen instead. In the first place, there are many cases where the problem hasn't yet been considered fully enough to establish a natural integration path. But in the second place, it would often be argued that nonspecialized calculations are more efficient because they produce ancillary results too. Furthermore, the success of alternative free energy calculations often depends on some physical insight. So those alternative approaches often have the virtue of testing a physical insight specific to the problem, and that would be a separate advantage counterbalancing the general numerical efficacy of thermodynamic integration.

5.1 Reference systems and umbrella sampling

We have several times above exploited knowledge of a physically relevant reference system to write

$$e^{-\beta \mu_\alpha^{ex}} = e^{-\beta \tilde{\mu}_\alpha^{ex}} \left\langle\!\left\langle e^{-\beta\left(\Delta U_\alpha - \Delta \tilde{U}_\alpha\right)}\right\rangle\!\right\rangle_r . \tag{5.1}$$

In this approach $e^{-\beta \Delta \tilde{U}_\alpha}$ is selected on the basis of physical insight, and the goal is to reduce the uncertainty of the estimate. This might permit simplified physical theories for the average.

Another view is that the additional factor $e^{\beta \Delta \tilde{U}_\alpha}$ serves to broaden the sampling. With this view, we might consider another configurational function $\Omega(\mathcal{N}+1)$ that helpfully broadens the sampling and write

$$e^{-\beta \mu_\alpha^{ex}(\mathcal{R}^n)} = \frac{\left\langle \left(e^{-\beta \Delta U_\alpha} \Omega\right) | \mathcal{R}^n \right\rangle_{P_B/\Omega}}{\left\langle \Omega^n | \mathcal{R}^n \right\rangle_{P_B/\Omega}}. \tag{5.2}$$

The sampling distribution indicated for $\langle \ldots | \mathcal{R}^n \rangle_{P_B/\Omega}$ is proportional to $P_B(\mathcal{N})/\Omega(\mathcal{N}+1)$. The denominator of Eq. (5.2) corresponds to the factor $e^{\beta \bar{\mu}_\alpha^{ex}(\mathcal{R}^n)}$ of Eq. (5.1). In contrast to the view of Eq. (5.1), here it is typically not assumed that $\langle \Omega | \mathcal{R}^n \rangle_{P_B/\Omega}$ is separately known. Thus bias or variance reduction issues would involve both the numerator and the denominator of Eq. (5.2).

Ω is called an *importance function* or sometimes an *umbrella function* (Torrie and Valleau, 1977). The latter name arises from the view that Ω broadens the sampling to cover relevant cases more effectively. Since Ω is involved as an unnormalized probability, it shouldn't change sign, and it shouldn't be zero throughout regions where the unmodified distribution and integrand are nonzero.

The *importance* language has a longer history than does *umbrella*; this reflects the fact that adjusting the sampling with *importance functions* is one of the small number of general tricks of Monte Carlo methods (Hammersley and Handscomb, 1964; Kalos and Whitlock, 1986). It is worth making this point more basically because this trick can be disguised. Consider a naive Monte Carlo estimate of the integral

$$I = \int\limits_0^1 f(x)\mathrm{d}x \approx \frac{1}{N} \sum_{i=1}^N f(x_i), \tag{5.3}$$

where the N sample points are drawn from a uniform distribution on the interval $0 \leq x \leq 1$. Importance sampling to improve the estimate exploits any knowledge of a normalized density $p(x)$ that might reduce the variation of the ratio $f(x)/p(x)$. Then

$$I = \int\limits_0^1 \left[\frac{f(x)}{p(x)}\right] p(x)\mathrm{d}x \approx \frac{1}{N} \sum_{i=1}^N \left[\frac{f(x_i)}{p(x_i)}\right], \tag{5.4}$$

where now the points are drawn from the distribution $p(x)$. The restrictions for $p(x)$ are essentially the same as those for Ω discussed above except that here we have assumed that $p(x)$ is normalized. A notational curiosity is that this can be written as

$$I = \int_0^1 f(x)e^{-\ln p(x)}dP \qquad (5.5)$$

with $dP = p(x)dx$ and $P(x)$ the cumulative distribution of $p(x)$. A change of integration coordinate has mapped the integration interval, $x = x(P)$, and $p(x)$ is the Jacobian for this coordinate change.

As an example of importance sampling ideas, consider the situation that the actual interest is in a family of solutes. When this comes up in pharmaceutical contexts, the family might be tens of molecules (Shaikh *et al.*, 2004). Is there a good reference system to use to get comparative thermodynamic properties for all members of this family? Let's simplify the question by supposing specific interest in a particular conformation of each molecule. There is a theoretical answer that is analogous to the Hebb training rule of neural networks (Hertz *et al.*, 1991; Plishke and Bergerson, 1994), and generalizes a procedure of Bennett (1976):

$$1/\Omega = \sum_{m\varepsilon\{\text{molecules}\}} e^{-\beta\Delta U_m} \qquad (5.6)$$

The sum is over all molecules in the family, each in the specific conformation of interest. When this reference potential is used to get the free energy for a specific solute in the family, it will match at least one contribution in the sum. So the umbrella covers everybody in the family, and this is literally the point of the original *umbrella sampling*: P_B/Ω "should cover simultaneously the regions of configuration space relevant to two or more physical systems" (Torrie and Valleau, 1977). Jointly matching several members of the family will help too. The penalty is just the sum over the family. The denominator of Eq. (5.2) is not germane to the evaluation of free energy differences between the members of the family.

Exercises

5.1 Consider reference interactions established as required by Eq. (5.1) and demonstrate the analogue of Eq. (3.25), p. 42:

$$P(\mathcal{N}+1) = \exp\{-\beta[\Phi - (\mu_\alpha^{\text{ex}} - \tilde{\mu}_\alpha^{\text{ex}})]\} \times \tilde{P}(\mathcal{N}+1), \qquad (5.7)$$

and

$$\mathcal{P}_\alpha(\varepsilon) = \exp\{-\beta[\varepsilon - (\mu_\alpha^{\text{ex}} - \tilde{\mu}_\alpha^{\text{ex}})]\} \tilde{\mathcal{P}}_\alpha(\varepsilon), \qquad (5.8)$$

where, for example, $\tilde{\mathcal{P}}_\alpha(\varepsilon)$ is defined by Eq. (4.5), p. 62 and $\mathcal{P}(\varepsilon)$ corresponds to fully coupled sampling of energies ε for interactions over-and-above the reference interactions.

5.2 Following the additive suggestion Eq. (5.6), consider the proposal

$$1/\Omega = e^{-\beta \Delta U_\alpha} + 1. \tag{5.9}$$

Intuitively, this corresponds to the possibilities that the distinguished molecule is present or not. Show that

$$e^{-\beta \mu_\alpha^{ex}} = \frac{\left\langle \frac{e^{-\beta \Delta U_\alpha/2}}{e^{-\beta \Delta U_\alpha/2}+e^{\beta \Delta U_\alpha/2}} \right\rangle_{P_B/\Omega}}{\left\langle \frac{e^{\beta \Delta U_\alpha/2}}{e^{-\beta \Delta U_\alpha/2}+e^{\beta \Delta U_\alpha/2}} \right\rangle_{P_B/\Omega}}. \tag{5.10}$$

5.3 Comparisons available from statistical data allow us to discuss a foundational issue, the possibility of *entropy calculated from the trajectory of motion*, raised years ago (Ma, 1985). This is a digression that illustrates the use of reference systems and statistical comparisons for the conceptualization of entropy.

The topic arises from the following sequence of aspects of entropy: when entropy is introduced on a thermodynamic basis the issue is the motion of heat (Jaynes, 1988), and the assessment involves calorimetry; an entropy change is evaluated. When entropy is formalized with the classical view of statistical thermodynamics, the entropy is found by evaluating a configurational integral (Bennett, 1976). But a macroscopic physical system at a particular thermodynamic state has a particular entropy, a state function, and the whole description of the physical system shouldn't involve more than a mechanical trajectory for the system in a stationary, equilibrium condition. How are these different concepts compatible?

The previous discussion (Ma, 1985) considered a lattice (Ising) model as a physical example and focused the concepts on recurrences in the trajectory. A method of evaluating the entropy was suggested. Here we carry that suggestion further, and discuss how that idea can be used to estimate values of integrals. We consider a classic quadrature problem, evaluation of an integral such as

$$I = \int\limits_0^1 f(x)\mathrm{d}x. \tag{5.11}$$

Figure 5.1 shows an example that can be kept in mind with this discussion. (As for many Monte Carlo methods exemplified with low-dimensional cases, sophisticated quadrature approaches are available here and would be much more efficient than Monte Carlo methods.)

Suppose that configurations visited by a thermal trajectory, for example from Markov Chain Monte Carlo, sample points uniformly in the allowed area. Let M be the total number of points involved and $m(\lambda)$ be the number

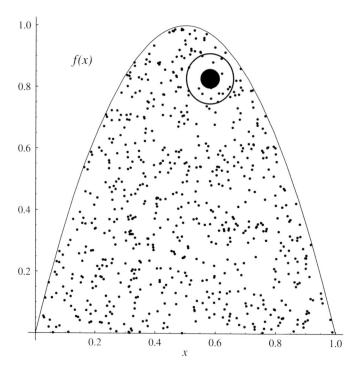

Figure 5.1 Example accompanying the discussion following Eq. (5.11). The function $f(x) = 4x(1 - x)$ is plotted. The dots indicate points drawn from a uniform distribution in the enclosed area. The bull's-eye is a distinguished point.

of points within a disk of small radius λ of a distinguished point. Explain why the combination

$$\pi \lambda^2 M / m(\lambda) \approx I \tag{5.12}$$

estimates the area under the curve when λ is sufficiently small. What might go wrong if λ were too large?

5.2 Overlap methods

With Eq. (3.25), p. 42, we have already indicated that the distribution function of the binding energy for the distinguished solute in the actual, fully-coupled system $\mathcal{P}_\alpha(\varepsilon) = \langle \delta(\varepsilon - \Delta U_\alpha) \rangle$ is related to the distribution function $\mathcal{P}_\alpha^{(0)}(\varepsilon)$ of Eq. (3.5), p. 33 by

$$\mathcal{P}_\alpha(\varepsilon) = e^{-\beta(\varepsilon - \mu_\alpha^{\mathrm{ex}})} \mathcal{P}_\alpha^{(0)}(\varepsilon). \tag{5.13}$$

This has something in common with the importance sampling discussed in the preceding section, except that the parameter sought, μ_α^{ex}, is now definitely supplying

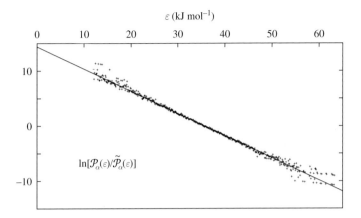

Figure 5.2 Assessment of electrostatic contributions to the excess chemical potential of water, following Eq. (5.15) redrawn from Hummer *et al.* (1995). The temperature is $T = 298\,K$ and the density is $\rho = 0.03333\,\text{Å}^{-3}$. The SPC model of water was used and the reference system interactions are those interactions with all partial charges given the value zero.

the normalization needed to implement the new sampling. Therefore, information on both distributions might be used to extract that thermodynamic quantity (Bennett, 1976; Ciccotti *et al.*, 1987)

$$\ln\left[\mathcal{P}_\alpha(\varepsilon)/\mathcal{P}_\alpha^{(0)}(\varepsilon)\right] = -\beta\varepsilon + \beta\mu_\alpha^{\text{ex}}. \tag{5.14}$$

Graphed as a function ε this should be linear with slope $-\beta$. The intercept then provides the thermodynamic parameter sought.

Similarly, if chemical potential contributions in excess of a defined reference system are sought, then

$$\ln\left[\mathcal{P}_\alpha(\varepsilon)/\tilde{\mathcal{P}}_\alpha(\varepsilon)\right] = -\beta\varepsilon + \beta\mu_\alpha^{\text{ex}} - \beta\tilde{\mu}_\alpha^{\text{ex}}, \tag{5.15}$$

following Eq. (5.8). p. 102.

5.3 Perturbation theory

To organize the description of interactions of a specified type, it is often helpful to introduce an ordering parameter λ in

$$\mu_\alpha^{\text{ex}}(\mathcal{R}^n) = \tilde{\mu}_\alpha^{\text{ex}}(\mathcal{R}^n) - kT\ln\left\langle e^{-\beta\lambda\Phi}|\mathcal{R}^n\right\rangle_r. \tag{5.16}$$

λ might also be viewed as a perturbative parameter in cases where it appears naturally as a gauge of the strength of solute–solvent interactions. In either case,

Table 5.1 *Successive cumulants,* C_j, *following Eq. (5.18).*
See also Table 6.1.

j	C_j
1	$\langle x \rangle$
2	$\langle x^2 \rangle - \langle x \rangle^2 = \langle \delta x^2 \rangle$
3	$\langle x^3 \rangle - 3\langle x \rangle \langle x^2 \rangle + 2\langle x \rangle^3 = \langle \delta x^3 \rangle$
4	$\langle x^4 \rangle - 4\langle x \rangle \langle x^3 \rangle + 12\langle x \rangle^2 \langle x^2 \rangle - 3\langle x^2 \rangle^2 - 6\langle x \rangle^4 = \langle \delta x^4 \rangle - 3\langle \delta x^2 \rangle^2$

the goal is to express $\beta \mu_\alpha^{\text{ex}}(\mathcal{R}^n)$ as a series ordered in powers of λ. Using the standard series for e^x and $\ln(1+x)$, and collecting powers of λ, we find

$$\mu_\alpha^{\text{ex}}(\mathcal{R}^n) = \tilde{\mu}_\alpha^{\text{ex}}(\mathcal{R}^n) + \langle \Phi | \mathcal{R}^n \rangle_{\text{r}} \lambda - \langle \delta \Phi^2 | \mathcal{R}^n \rangle_{\text{r}} \frac{\beta \lambda^2}{2} \cdots \qquad (5.17)$$

where $\delta \Phi = \Phi - \langle \Phi | \mathcal{R}^n \rangle_{\text{r}}$. This result should be compared with Eqs. (4.8), p. 64 and (4.12), p. 66 that we obtained on the basis of gaussian distribution models. Note again that the final term of Eq. (5.17) necessarily lowers the value of the excess chemical potential.

This expansion in powers of λ can be viewed from a more general perspective. The counting and arranging of powers of λ is a formal operation, so we can carry out that analysis on the basis of a simpler notation, e.g.

$$\ln \left\langle e^{\lambda x} \right\rangle = \sum_{j=1} C_j \frac{\lambda^j}{j!}, \qquad (5.18)$$

where x is the random variable, and the coefficients C_j are *cumulants* (Kubo, 1962); sometimes in the older literature they are referred to as the semi-invariants of Thiele (Graham *et al.*, 1994). The calculation leading to Eq. (5.17) establishes the first two cumulants, and the next two cumulants after that are given in Table 5.1. More coefficients than those four are rarely useful because convergence is typically problematic: cumulant series truncated beyond second order are not necessarily realizable (Marcienkeiwicz, 1939). If an additional term or two beyond the second order – gaussian – theory doesn't provide a minor improvement of a generally satisfactory result, then some other approach is probably required.

Thermodynamic perturbation theory

Thermodynamic perturbation theory has an extended history (Peierls, 1933; Kirkwood, 1983; Zwanzig, 1954; Landau *et al.*, 1980, Section 3.2; Peierls, 1979, Section 3.3) and a decided practical utility. Here we discuss one of the subtler points. The common case for thermodynamic perturbation theory is to focus on the difference of the Helmholtz free energies between two systems with interactions described by $U_I(\mathcal{N})$ and $U_{II}(\mathcal{N})$ (McQuarrie, 1976):

$$\exp\left[-\beta\left(\mathcal{A}_{II} - \mathcal{A}_I\right)\right] = \left\langle\exp\left[-\beta\left(U_{II}(\mathcal{N}) - U_I(\mathcal{N})\right)\right]\right\rangle_I. \qquad (5.19)$$

The indicated averaging $\langle\ldots\rangle_I$ is obtained for the system with interactions $U_I(\mathcal{N})$. Equation (5.19) has a strong formal similarity to Eq. (5.1). Beyond the formalities, however, there is a subtle difference that is important. The system described by the interactions $U_I(\mathcal{N})$ could be qualitatively different from that corresponding to interactions $U_{II}(\mathcal{N})$, even if the differences in interactions seem small on a molecular basis: for example, the two systems could even adopt different thermodynamic phases! This is a symptom of the fact that the energy differences exponentiated in Eq. (5.19) are *extensive* in the system size. This difficulty won't arise with Eq. (5.1) as we have already emphasized. On a more workmanlike basis, exponentiating energy differences that are extensively large will promptly make practical calculations unmanageable. It is a striking point that the initial attempts to use Eq. (5.19) (Torrie and Valleau, 1977; Valleau and Torrie, 1977) in practical calculations were limited to unusually small systems and to small changes in the interactions.

We can begin to address the issue of extensivity by further theoretical development, rewriting Eq. (5.19) as

$$e^{-\beta\Delta\mathcal{A}} = \int w(N\varepsilon)e^{-N\beta\varepsilon}\,\mathrm{d}\,(N\varepsilon) \qquad (5.20)$$

with

$$w(N\varepsilon) \equiv \left\langle\delta\left(N\varepsilon - (U_{II}(\mathcal{N}) - U_I(\mathcal{N}))\right)\right\rangle_I. \qquad (5.21)$$

In the usual thermodynamic setting, we expect that $\ln w(N\varepsilon) \sim O(N)$ because distant regions of a macroscopically large system will be practically uncorrelated. And we expect $\ln w(N\varepsilon)$ to increase with increasing $N\varepsilon$, i.e. there are more states available at higher energy. Therefore, we expect to be able to find $N\varepsilon = N\bar{\varepsilon}$ at which point the exponents of the integrand of Eq. (5.20) will balance, and we write

$$\ln\left[w(N\varepsilon)e^{-N\beta\varepsilon}\right] \approx -N\beta\bar{\varepsilon} + \ln w(N\bar{\varepsilon})$$

$$+ (\varepsilon - \bar{\varepsilon})\left[\frac{\mathrm{d}\ln w(N\bar{\varepsilon})}{\mathrm{d}\bar{\varepsilon}} - N\beta\right]$$

$$+ \frac{(\varepsilon - \bar{\varepsilon})^2}{2}\frac{\mathrm{d}^2\ln w(N\bar{\varepsilon})}{\mathrm{d}\bar{\varepsilon}^2}. \qquad (5.22)$$

Locating the maximum of the integrand by

$$\frac{d \ln w(N\bar{\varepsilon})}{dN\bar{\varepsilon}} = \beta, \tag{5.23}$$

and using the gaussian approximation, to leading order in N, we find

$$\Delta \mathcal{A} \sim N\bar{\varepsilon} - kT \ln w(N\bar{\varepsilon}). \tag{5.24}$$

Together with Eq. (5.24), Eq. (5.23) suggests the thermodynamic verity $\partial S/\partial E = 1/T$. Note also the expected behavior that the coefficient of the quadratic term Eq. (5.22) scales with $1/N$, and the variance associated with the gaussian integrand is $kT^2 C_V / N^2$. The conclusion of this argument is that thermodynamic perturbation theory as addressed generally through Eq. (5.19) will depend on a small fraction of the statistical data in the neighborhood of $N\varepsilon = N\bar{\varepsilon}$, a fraction which vanishes exponentially with N in the large system limit; and that will be true even when the physical differences between $U_{\mathrm{I}}(\mathcal{N})$ and $U_{\mathrm{II}}(\mathcal{N})$ are minor. This issue is important for practical calculations but needn't be a problem for calculations of a more theoretical type. Still, analytical expression of the second-order perturbation contribution to $\Delta \mathcal{A}$ requires special technical subtlety exactly because of this issue: because an extensive quantity is being considered in the exponent, weak correlations between distant molecule pairs have to be considered carefully (Zwanzig, 1954; Henderson and Barker, 1971; Hansen and McDonald, 1976, see Section 6.2).

Exercises

5.4 Evaluate the average on the left of Eq. (5.18) assuming a gaussian distribution for the random variable x, thereby establishing that the cumulant series truncated at second order is correct for a gaussian distribution.

5.5 Consider Eq. (5.60), p. 118, from the point of view of a cumulant expansion and derive

$$\mu_\alpha^{\mathrm{ex}}(\mathcal{R}^n) \approx \tilde{\mu}_\alpha^{\mathrm{ex}}(\mathcal{R}^n)$$

$$+ \frac{1}{M} \sum_{j=1}^{M} \left[\langle \Phi_\alpha | \mathcal{R}^n \rangle_{\lambda_j} + \frac{\beta^2}{24M^2} \langle \delta \Phi_\alpha^3 | \mathcal{R}^n \rangle_{\lambda_j} \right] \tag{5.25}$$

as an improvement to Eq. (5.57), p. 117. Conclude that the quadrature Eq. (5.57) is locally correct through the order of the second derivative of μ^{ex}.

Electrostatics in simulation: periodic boundary conditions

Intermolecular electrostatic interactions are often the target of perturbation analysis on the basis of simulation data (Hummer *et al.*, 1998*b*). We give here a non-traditional discussion of the Ewald electrostatic potential often used in computer simulation with periodic, or Born–von Karman, boundary conditions (Ashcroft and Mermin, 1976; Peierls, 1979, Section 3.6). In particular, we discuss how Poisson's equation can be solved in periodic boundary conditions, avoiding the delicacies of the conditionally convergent lattice sums (Leeuw *et al.*, 1980). Ideas developed in several exercises are then employed to consider the Ewald potential appropriate for non-neutral solutes to obtain size-consistent results for small systems (Hummer *et al.*, 1998*b*). This approach is useful in perturbation theories of ion solvation. The outlook of this extended discussion is to encourage the theoretical analysis of the Ewald potential for its physical consequences, in addition to the analysis of the computational exertion of evaluating it.

We begin by considering a neutral system confined to a simulation cell. We seek to solve the Poisson equation

$$\nabla^2 \Phi(r) = -4\pi \varrho(r), \tag{5.26}$$

where $\varrho(r)$ is the density of electric charge, under the condition that the solution should be periodic with the simulation cell as the fundamental period. Because of the linearity of Eq. (5.26), we will separate contributions from elementary charge sources, and denote those contributions by $\varphi(r)$. This step is less innocuous than it sounds, and we will discuss the consequences in what follows. Because of the intended periodicity, when considering an elementary charge source we translate the simulation cell so that the source is at the center. It is then obvious that the surface integral

$$\int_{\text{cell}} \hat{n} \cdot \nabla \varphi(r) \mathrm{d}^2 r \tag{5.27}$$

for the electric potential $\varphi(r)$ due to that elementary charge source should be zero because of the periodicity. But the straightforward Gauss's law calculation shows this to be impossible if the net charge enclosed isn't zero. A simple modification that permits periodic solutions for the separated contributions is obtained by inserting a uniform neutralizing background; thus

$$\nabla^2 \varphi(r) = -4\pi \left(\delta(r) - \frac{1}{V} \right) \tag{5.28}$$

for a unit magnitude source. If the net charge of the system is zero, this modification is innocuous because the sum of the background contributions to the elementary charges vanishes. If the net charge of the system isn't zero, then this is a reasonable modification of the problem to permit a periodic solution.

To proceed further, we consider the simplification of a cubic cell so that $\hat{n} \cdot \nabla\varphi = 0$ identically on the surface of the cell, and write the usual Green's theorem application: see Eq. (5.31) below or Jackson (1975)

$$\varphi(\boldsymbol{r}) = \frac{1}{r} - \frac{1}{\mathcal{L}^3} \int\limits_{\text{cell}} \frac{\text{d}^3 r'}{|\boldsymbol{r} - \boldsymbol{r}'|} - \frac{1}{4\pi} \int\limits_{\text{surface}} \hat{n} \cdot \nabla' \left(\frac{1}{|\boldsymbol{r} - \boldsymbol{r}'|} \right) \varphi(\boldsymbol{r}') \text{d}^2 r'. \qquad (5.29)$$

The first two terms on the right comprise the Coulomb potential due to the source and the neutralizing background. The last term on the right is a dipole layer contribution from the surface, necessary to achieve $\hat{n} \cdot \nabla\varphi = 0$ on the surface.

Solution of the Poisson equation in periodic boundaries by finite difference methods (Beck, 2000) provides yet another way to view the Ewald potential. In that case, iterative methods with the periodic boundary conditions converge provided the net charge is zero inside the domain. The charge neutrality is enforced by addition of a uniform neutralizing background. Each iterative step involves only near-local information in the neighborhood of a given grid point, and potential values are only required within the volume $\mathcal{V}$ and on the boundaries. The solution can be made unique by imposing a subsidiary condition such as setting the integral of the potential over the domain to zero.

The alternative viewpoints here emphasize that the uniform neutralizing background for the individual contributions just permits the normal electric field to be zero on the boundary. These viewpoints avoid traditional (Valleau and Torrie, 1977) but inconclusive discussions of what periodic images might be doing when lattice sums are conceived with Ewald potentials.

Exercises

5.6 Consider solutions of the Poisson equation in one spatial dimension for a system with a uniform charge density $1/\mathcal{L}$ in the interval $-\mathcal{L}/2 < x < \mathcal{L}/2$. Write out the general solution for this domain, and discuss the possibilities for making this general solution periodic with period $\mathcal{L}$.

5.7 Evaluate the derivatives of

$$\varphi(x) = -2\pi |x| + 2\pi \int\limits_{-\mathcal{L}/2}^{\mathcal{L}/2} \frac{|x - x'|}{\mathcal{L}} \text{d}x' \qquad (5.30)$$

to compose the Poisson equation on $-\mathcal{L}/2 < x < \mathcal{L}/2$. What is the charge density implied?

5.8 Starting from

$$\phi \frac{\text{d}^2 \psi}{\text{d}x^2} = \frac{\text{d}}{\text{d}x} \left(\phi \frac{\text{d}\psi}{\text{d}x} \right) - \frac{\text{d}\psi}{\text{d}x} \frac{\text{d}\phi}{\text{d}x} \qquad (5.31)$$

construct Eq. (5.30) as the one-dimensional analog of the Green's theorem result Eq. (5.29).

5.9 From Eq. (5.30) show that

$$\varphi(x) = -2\pi \left(|x| - \frac{x^2}{\mathcal{L}} \right), \tag{5.32}$$

achieves $\varphi'(\mathcal{L}/2) = 0$, and being an even function of x can be a periodic solution of Eq. (5.28) in one dimension.

5.10 The Ewald potential traditionally includes an additive constant to achieve

$$\int_{\text{cell}} \varphi(\boldsymbol{r}) \mathrm{d}^3 r = 0. \tag{5.33}$$

Since periodic boundary conditions preserve translational homogeneity, a physical perspective on this requirement is that the *potential of the phase –* see p. 69 – is zero. Show that the potential

$$\varphi(x) = -2\pi |x| + \frac{2\pi}{\mathcal{L}} (x^2 + \mathcal{L}^2/6), \tag{5.34}$$

agrees with this requirement for the one-dimensional case. See Fig. 5.3.

5.11 Show that the solution of Eq. (5.29) is not otherwise changed by adjusting $\varphi(\boldsymbol{r})$ by a spatially uniform constant. Hint: remember that

$$\int_{\text{surface}} \hat{\boldsymbol{n}} \cdot \nabla' \left(\frac{1}{|\boldsymbol{r} - \boldsymbol{r}'|} \right) \mathrm{d}^2 r' = -4\pi. \tag{5.35}$$

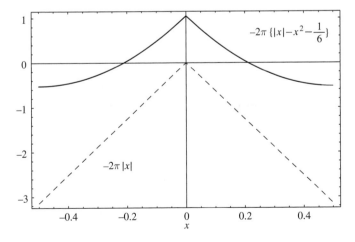

Figure 5.3 Comparison of Ewald and bare electrostatic potential in one dimension. See Eq. (5.32).

5.12 Equation (5.29) is obtained from a standard Green's theorem expression (Jackson, 1975) by the identification

$$\int_{\text{surface}} \frac{\hat{\boldsymbol{n}} \cdot \nabla' \varphi(\boldsymbol{r})}{|\boldsymbol{r} - \boldsymbol{r}'|} d^2 r' = 0 \tag{5.36}$$

following from the intention that the normal derivative should be identically zero. Consider the converse issue: does a solution of Eq. (5.29) necessarily imply that the normal derivative should be identically zero? Hint: $\hat{\boldsymbol{n}} \cdot \nabla' \varphi(\boldsymbol{r})$ functions in Eq. (5.29) as a surface charge density, so a Gauss's law calculation might be useful.

The Ewald potential and lattice sums

The Ewald potential is traditionally implemented as a lattice sum (Ziman, 1972; Leeuw *et al.*, 1980). We just outlined a conceptualization of electrostatic interactions in periodic boundary conditions that involved adding a uniform neutralizing background for each charge, and the subsequent solution of the Poisson equation in periodic boundary conditions. Here we discuss the interconnections between that conceptualization and the traditional lattice sums, as presented in many sources, e.g. (Allen and Tildesley, 1987; Frenkel and Smit, 2002; Leeuw *et al.*, 1980).

Consider a system composed of a collection of molecules and ions, with partial charges distributed according to the details of the force field, in a cubic volume $\mathcal{V} = \mathcal{L}^3$. Let's begin by considering the electrostatic potential

$$\psi(\boldsymbol{r}) = \sum_{\boldsymbol{n}} \frac{\text{erfc}(\eta|\boldsymbol{r} + \boldsymbol{n}\mathcal{L}|)}{|\boldsymbol{r} + \boldsymbol{n}\mathcal{L}|} + \frac{4\pi}{\mathcal{L}^3} \sum_{\boldsymbol{k} \neq 0} \left(\frac{e^{-k^2/4\eta^2}}{k^2} \right) e^{i\boldsymbol{k}\cdot\boldsymbol{r}} - \frac{\pi}{\mathcal{L}^3 \eta^2} \tag{5.37}$$

where $\boldsymbol{n} = (n_x, n_y, n_z)$ is a vector with integer components, $\boldsymbol{k} = \frac{2\pi}{\mathcal{L}}\boldsymbol{n}$, and

$$\text{erfc}(z) \equiv 1 - \frac{2}{\sqrt{\pi}} \int_0^z e^{-u^2} du. \tag{5.38}$$

An exercise that follows requests utilization of the Poisson equation to evaluate the charge density implied by Eq. (5.37); the result of that calculation is the charge density in the Poisson equation of Eq. (5.26), p. 109. The potential in Eq. (5.37) is identical to traditional expressions for the Ewald potential (Leeuw *et al.*, 1980) except for the final term; that additive constant implies

$$\int_{\text{cell}} \psi(\boldsymbol{r}) d\boldsymbol{r} = 0 \tag{5.39}$$

because

$$\int\limits_{\text{cell}} \sum_n \frac{\text{erfc}(\eta|r+n\mathcal{L}|)}{|r+n\mathcal{L}|} dr = \frac{\pi}{\eta^2}. \qquad (5.40)$$

The spatial integral of the middle term of Eq. (5.37) is zero because of the $k \neq 0$ exclusion. In addition, the potential $\psi(\mathbf{r})$ has the virtue of being independent of η. In view of Eq. (5.39), the potential energy

$$\frac{1}{2} \int\limits_{\text{cell}} \left[\sum_j q_j \left(\delta(r-r_j) - \frac{1}{\mathcal{L}^3} \right) \right] \sum_k q_k \psi(r-r_k) dr$$

$$= \frac{1}{2} \sum_{jk} q_j q_k \psi(r_{kj}) \qquad (5.41)$$

won't specifically exhibit the background charge. This potential energy, however, formally includes a bare self-interaction that we wish to exclude. Separating out the $j = k$ contribution, and then excluding the bare self-interaction produces

$$U = \frac{1}{2} \sum_{j \neq k} q_j q_k \psi(r_{kj}) + \frac{1}{2} \sum_k q_k^2 \lim_{r \to 0} \left(\psi(r) - \frac{1}{r} \right)$$

$$\equiv \frac{1}{2} \sum_{j \neq k} q_j q_k \psi(r_{kj}) + \frac{\xi}{2\mathcal{L}} \sum_k q_k^2. \qquad (5.42)$$

This energy includes all the relevant interactions after removing the interaction of the ion in the cell with itself; the interaction of the ion with its periodic images is retained. The potential $\xi/\mathcal{L}$ is that due to the neutralizing background in the cell and all periodic images of the unit charge and the background. From Eq. (5.37), we see that this potential is given by

$$\frac{\xi}{\mathcal{L}} = \sum_{n \neq 0} \frac{\text{erfc}(\eta|n\mathcal{L}|)}{|n\mathcal{L}|} + \frac{4\pi}{\mathcal{L}^3} \sum_{k \neq 0} \left(\frac{e^{-k^2/4\eta^2}}{k^2} \right) - \frac{\pi}{\mathcal{L}^3 \eta^2} - \frac{2\eta}{\sqrt{\pi}}, \qquad (5.43)$$

which has used the relation

$$\lim_{r \to 0} \left(\frac{\text{erfc}(\eta r)}{r} - \frac{1}{r} \right) = -\frac{2\eta}{\sqrt{\pi}}. \qquad (5.44)$$

Thus after exclusion of the bare interaction from the $n = 0$ term of the real-space sum, that contribution yields a constant. In practical calculations it is common to use a convergence parameter of $\eta \approx 5.6/\mathcal{L}$, in which case the first term on the right side of Eq. (5.43) is negligibly small, and we shall omit this term in the discussion to follow. The numerical value of ξ for a cubic lattice is approximately -2.837297 (see the Exercises). The choice of Eq. (5.39) thus removes the interaction of each ion with the charge backgrounds in this energy expression, but a self-interaction remains. The displayed self-energy is merely a constant for the common case of permanent particles with permanent charges. On the other hand, if the calculation manipulates the charge state of solution species, as in Section 4.2, p. 67, then specific awareness of this self-energy has an advantage of mechanical consistency. Figure 5.4 displays the distribution of electrostatic interaction energies for an ion in water with and without the above-discussed correction terms. This illustrates the importance of these corrections in obtaining reliable hydration free energies.

The textbook formula for the total potential energy of a collection of charges in the traditional Ewald method (Allen and Tildesley, 1987; Frenkel and Smit, 2002) is:

$$U = \frac{1}{2} \sum_{i \neq j}^{N} q_i q_j \left\{ \sum_{n} \frac{\mathrm{erfc}(\eta |\mathbf{r}_{ij} + \mathbf{n}\mathcal{L}|)}{|\mathbf{r}_{ij} + \mathbf{n}\mathcal{L}|} \right\}$$

$$+ \frac{1}{2} \sum_{i,j}^{N} q_i q_j \left\{ \frac{4\pi}{\mathcal{L}^3} \sum_{k \neq 0} \left(\frac{e^{-k^2/4\eta^2}}{k^2} \right) e^{i\mathbf{k}\cdot\mathbf{r}_{ij}} \right\} - \frac{\eta}{\sqrt{\pi}} \sum_{i}^{N} q_i^2. \qquad (5.45)$$

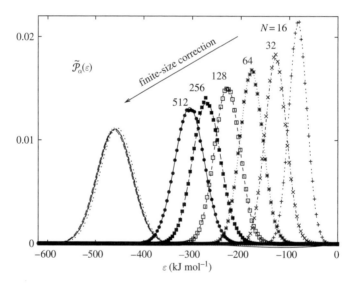

Figure 5.4 Finite-size correction of the probability densities of the electrostatic energies of positively charged imidazolium ion in water. The uncorrected distributions are shown with symbols, together with corresponding Gaussian distributions. In addition to the electrostatic correction, a thermodynamic correction is also applied, but this correction is small in magnitude, see Hummer *et al.* (1998*b*). With the corrections, the distributions collapse and agree closely for all system sizes of $16 \leq N \leq 512$ water molecules.

Charge balance is assumed in the derivation of this energy formula (Ziman, 1972). A following exercise calculates the total potential energy at an ion using Eqs. (5.42) and (5.45); those results show that the traditional treatment (Allen and Tildesley, 1987; Frenkel and Smit, 2002; Leeuw *et al.*, 1980), associated with conducting boundary conditions, is obtained for the case of charge-neutral systems.

Exercises

5.13 Evaluate the Poisson equation for the potential of Eq. (5.37) to determine the charge density implied. Hints: relationships developed in Arfken (1985) are helpful. Additionally, the *theta-function transformation* (Ziman, 1972)

$$\frac{4\eta^3}{\sqrt{\pi}} \sum_n e^{-\eta^2 |r - n\mathcal{L}|^2} = \frac{4\pi}{\mathcal{L}^3} \sum_k e^{-k^2/4\eta^2} e^{ik \cdot r} \tag{5.46}$$

might be useful.

5.14 Show that $\partial\psi(r)/\partial\eta = 0$. See the previous exercise for useful information.

5.15 Prove Eq. (5.40).

5.16 Write a computer program to compute ξ using Eq. (5.43). Can you develop an approximate analytical expression that yields a similar result?

5.17 Prove Eq. (5.44).

5.18 Consider η^{-1} as a length scale parameter, and develop a dimensional scaling argument to conclude that

$$\frac{4\pi}{\mathcal{L}^3} \sum_{k \neq 0} \frac{e^{-k^2/4\eta^2}}{k^2} \sim \eta f(\eta\mathcal{L}). \tag{5.47}$$

Show that

$$\lim_{\mathcal{L} \to \infty} \frac{4\pi}{\mathcal{L}^3} \sum_{k \neq 0} \frac{e^{-k^2/4\eta^2}}{k^2} = \frac{2\eta}{\sqrt{\pi}}. \tag{5.48}$$

This is the infinite-volume limit of the corresponding sum obtained from Eq. (5.43).

5.19 Show that

$$\sum_{i \neq j} q_i q_j = Q^2 - \sum_i q_i^2 \tag{5.49}$$

where Q is the total charge of the system.

5.20 Determine the total electrostatic potential at an ion site from the two expressions Eqs. (5.42) and (5.45). The potential derived from Eq. (5.45) is appropriate for a system with overall charge neutrality. Note that the k-space sum in Eq. (5.45) includes all i, j terms as opposed to Eq. (5.42). The total electrostatic energy is simply $U = 1/2 \sum_i q_i \phi_i^{tot}(r_i)$.

5.21 Using the above result, show that the total potential at an ion for a charged system compensated by a neutralizing background is

$$
\phi^{\text{tot}}(\mathbf{r}_i) = \sum_{i\neq j}^{N} q_j \left\{ \sum_{n} \frac{\operatorname{erfc}(\eta|\mathbf{r}_{ij}+\mathbf{n}\mathcal{L}|)}{|\mathbf{r}_{ij}+\mathbf{n}\mathcal{L}|} \right\}
$$
$$
+ \sum_{j}^{N} q_j \left\{ \left(\frac{4\pi}{\mathcal{L}^3}\right) \sum_{k\neq 0} e^{i\mathbf{k}\cdot\mathbf{r}_{ij}} \frac{e^{-k^2/4\eta^2}}{k^2} \right\} - \frac{\pi}{\mathcal{L}^3\eta^2}Q - \frac{2\eta}{\sqrt{\pi}}q_i, \quad (5.50)
$$

which verifies that the traditional total potential is obtained for the case $Q=0$. This exercise shows that a simple modification of a traditional Ewald code can be used to construct the total potential for a charged system.

5.22 Show that if the total system charge Q is the charge q on one ion, the change in system total energy upon mutation of that charge from q_0 to q_1 is

$$
\Delta U = \Delta q\hat{\phi}(\mathbf{r}_i) + \frac{\xi}{2\mathcal{L}}(q_1^2 - q_0^2) \qquad (5.51)
$$

where $\Delta q = q_1 - q_0$ and

$$
\hat{\phi}(\mathbf{r}_i) = \sum_{j\neq i}^{N} q_j \left\{ \sum_{n} \frac{\operatorname{erfc}(\eta|\mathbf{r}_{ij}+\mathbf{n}\mathcal{L}|)}{|\mathbf{r}_{ij}+\mathbf{n}\mathcal{L}|} \right\}
$$
$$
+ \sum_{j\neq i}^{N} q_j \left\{ \left(\frac{4\pi}{\mathcal{L}^3}\right) \sum_{k\neq 0} e^{i\mathbf{k}\cdot\mathbf{r}_{ij}} \frac{e^{-k^2/4\eta^2}}{k^2} \right\}. \qquad (5.52)
$$

5.23 Use the previous result to show that the second-order cumulant expansion for the change in excess chemical potential is then

$$
\Delta\mu^{\text{ex}} = \Delta q\left[\langle\hat{\phi}\rangle_{q_0} + \frac{q_0\xi}{\mathcal{L}} \right] - \frac{\beta}{2}\Delta q^2\left[\langle(\hat{\phi}-\langle\hat{\phi}\rangle_{q_0})^2\rangle_{q_0} - \frac{\xi}{\beta\mathcal{L}} \right] + \cdots \quad (5.53)
$$

5.24 Extend the above discussion to a molecular ion with charges distributed at several sites on the ion. See Hummer *et al.* (1998*b*).

5.4 Thermodynamic integration

The derivative $\partial\mu_{\alpha}^{\text{ex}}(\lambda)/\partial\lambda = \left\langle\Delta U_{\alpha}^{(1)}\right\rangle_{\lambda}$ can be obtained straightforwardly from simulation data. On this basis, mere quadrature (Press *et al.*, 1992) provides an evaluation of

$$
\mu_{\alpha}^{\text{ex}} = \int_{0}^{1} \left\langle\Delta U_{\alpha}^{(1)}\right\rangle_{\lambda} \, d\lambda. \qquad (5.54)
$$

This is conceptually simple. But it would typically be a problematic approach for undifferentiated intermolecular interactions because that would typically include excluded volume interactions – reference system interactions in a van der Waals view – which can be infinitely large, and therefore highly variable. In such a case, the statistical efficiency, and perhaps also the statistical quality, is likely to be troublesome.

Instead the most natural use of thermodynamic integration is to treat interactions beyond physically defined reference interactions. To proceed in that direction, we begin from Eq. (4.4), p. 62, in the form

$$\mu_\alpha^{\text{ex}}(\mathcal{R}^n) = \tilde{\mu}_\alpha^{\text{ex}}(\mathcal{R}^n) + \int_0^1 \langle \Phi_\alpha | \mathcal{R}^n \rangle_\lambda \, d\lambda, \tag{5.55}$$

including the conformational coordinate for generality. The use of λ to scale partial atomic charges on the distinguished solute is an example. In such physically defined cases, the statistical quality of the derivative estimates are likely to be better controlled.

Exercises

5.25 Consider the case that a distinguished solute is coupled to the solution on the basis of the interaction $\Delta \tilde{U}_\alpha + \lambda \Phi_\alpha$. Show that

$$\frac{\partial \mu_\alpha^{\text{ex}}(\mathcal{R}^n)}{\partial \lambda} = \langle \Phi_\alpha | \mathcal{R}^n \rangle_\lambda, \tag{5.56a}$$

$$\frac{\partial^2 \mu_\alpha^{\text{ex}}(\mathcal{R}^n)}{\partial \lambda^2} = -\beta \langle \delta \Phi_\alpha^2 | \mathcal{R}^n \rangle_\lambda, \tag{5.56b}$$

where

$$\langle \delta \Phi_\alpha^2 | \mathcal{R}^n \rangle_\lambda = \langle \Phi_\alpha^2 | \mathcal{R}^n \rangle_\lambda - \langle \Phi_\alpha | \mathcal{R}^n \rangle_\lambda^2.$$

5.26 Notice that the second derivative Eq. (5.56b) is never positive. Use this observation to prove that $\mu_\alpha^{\text{ex}}(\mathcal{R}^n) - \tilde{\mu}_\alpha^{\text{ex}}(\mathcal{R}^n) \leq \langle \Phi_\alpha | \mathcal{R}^n \rangle_0$.

5.27 Consider implementing Eq. (5.55) on the basis of the mid-point rule and evaluation of the integral at M equally spaced points, $\lambda_j = \frac{(j-1/2)}{M}$:

$$\mu^{\text{ex}}(\mathcal{R}^n) - \tilde{\mu}^{\text{ex}}(\mathcal{R}^n) \approx \frac{1}{M} \sum_{j=1}^M \langle \Phi_\alpha | \mathcal{R}^n \rangle_{\lambda_j}. \tag{5.57}$$

Suppose that the variance of the estimate of each $\langle \Phi_\alpha | \mathcal{R}^n \rangle_{\lambda_j}$ is

$$\frac{1}{m} \langle \delta \Phi_\alpha^2 | \mathcal{R}^n \rangle_{\lambda_j}, \tag{5.58}$$

i.e. assume that there are m effectively uncorrelated blocks of data for the estimation of each summand of Eq. (5.57). Consider the statistical uncertainty of the estimate Eq. (5.57) and show that for large M the standard error σ is given by

$$\sigma^2(\mathcal{R}^n) \approx \frac{kT}{mM}[\langle\Phi_\alpha|\mathcal{R}^n\rangle_{\lambda=0} - \langle\Phi_\alpha|\mathcal{R}^n\rangle_{\lambda=1}]. \tag{5.59}$$

This can be used with literature results for thermodynamic integration that aren't accompanied by a specific statistical uncertainty.

5.28 Show that

$$\mu_\alpha^{ex}(\mathcal{R}^n) = \tilde{\mu}_\alpha^{ex}(\mathcal{R}^n) - kT\sum_{j=1}^{M}\ln\left[\frac{\langle e^{-\beta\Phi_\alpha/2M}\rangle_{\lambda_j}}{\langle e^{\beta\Phi_\alpha/2M}\rangle_{\lambda_j}}\right], \tag{5.60}$$

formally without quadrature error, as compared with Eq. (5.57).

5.5 Bias

We will introduce this subject through discussion of a central example. Suppose that we have a sample $\{\omega_1, \omega_2, \ldots, \omega_M\} = \Omega$ from which we will estimate a partition function $\mathcal{Z}$. This means that we have a function $\hat{Z}(\Omega)$ to be evaluated for our sample to produce a numerical value that we take to be the estimate of $\mathcal{Z}$. We assume here the common circumstance that $\hat{Z}(\Omega)$ takes the form

$$\hat{Z}(\Omega) = \frac{1}{M}\sum_{i=1}^{M}f(\omega_i). \tag{5.61}$$

We will also assume the common circumstance that $\hat{Z}(\Omega)$ is an *unbiased* estimator of $\mathcal{Z}$; in practical terms, this means that if the same estimation is performed repeatedly then the average of the estimates is the quantity desired, $\mathcal{Z} = E[\hat{Z}(\Omega)]$, the expected value of the estimator. The alternative notation for expectation is used here to avoid potential confusion with the statistical mechanical averages used throughout this book.

With this setup, notice what happens when we inquire about a value for a free energy $\beta F \equiv -\ln\mathcal{Z}$. Then

$$\beta\hat{F} = -\ln\hat{Z} = -\ln\left[\mathcal{Z} + \frac{1}{M}\sum_{i=1}^{M}(f(\omega_i) - \mathcal{Z})\right]. \tag{5.62}$$

Expanding to the lowest-order nonvanishing term, averaging the result obtained, and assuming that the sample points are *independent* and *identically distributed*, we obtain

$$E\left[\beta(\hat{F} - F)\right] \approx \frac{1}{2M}E\left[\left(\frac{f(\omega) - \mathcal{Z}}{\mathcal{Z}}\right)^2\right] > 0. \tag{5.63}$$

The difference $E[\hat{F} - F]$ is called the *bias*, and Eq. (5.63) is an approximate evaluation of the bias. This approximate bias is positive and diminishes proportionally to $1/M$ with increasing sample size. The conclusion of Eq. (5.63) is that the estimated free energy, utilizing a sample of size M, is greater than the true free energy; performing the same estimation many times doesn't change this.

The most straightforward remedy for this specific situation is clear. If we were to perform this estimation several times – perhaps m times – we would have several samples $\Omega^{(j)}, j = 1, \ldots, m$ of size M. We could aggregate these samples to produce one sample of size mM, evaluate $\hat{Z}$ with the aggregated sample, and the bias in the reported free energy should be smaller according to Eq. (5.63). Extrapolation of the estimates obtained from different possibilities for aggregating the samples is the basic idea behind the *jackknife* method of estimating the bias (Efron and Tibshirani, 1993; Zuckerman and Woolf, 2002). Generalization of these concepts leads to *bootstrap* methods for these data analysis issues (Efron and Tibshirani, 1993).

Exercises

5.29 *Jackknife estimate of the bias.* For the estimate discussed above, consider the jackknifed samples obtained by deleting one point from an original sample, $\Omega_{(k)} = \Omega - \omega_k$. Let $\hat{F}_{(k)}$ be the estimated free energy obtained with the kth jackknifed sample $\Omega_{(k)}$ so that

$$\beta\hat{F}_{(k)} = -\ln\left[\frac{1}{M-1}\sum_{i \neq k} f(\omega_i)\right] \equiv -\ln\hat{Z}_{(k)}. \qquad (5.64)$$

Let $\hat{F}_M = -\ln\hat{Z}_M$ be the estimate with the whole, original sample Ω. Using the approximate form Eq. (5.63), find the more evolved estimate:

$$\beta\hat{F} \approx \frac{1}{M}\sum_{i=1}^{M}\beta\hat{F}_{(i)} + \sum_{i=1}^{M}\left(\beta\hat{F}_M - \beta\hat{F}_{(i)}\right). \qquad (5.65)$$

5.30 The difference $\hat{Z}_M - \hat{Z}_{(k)}$ involves the contribution of only one observation, a small difference when M is large. Expand the last term of Eq. (5.65) to quadratic order in these small differences and compare the result with Eq. (5.63).

5.31 Reconsider the previous discussion of *entropy calculated from the trajectory of motion*, associated with Fig. 5.1, p. 104. A view taken in that previous exercise is that a large, fixed number M of data points is sampled uniformly in the area that is to be estimated. With a fixed reference subarea $\pi\lambda^2$, the number of points, $m(\lambda)$, within the reference subarea is a random variable.

Observations of $m(\lambda)$ are used to estimate the total area. An alternative view would be to adopt a target value m and find the $\lambda(m)$ that encloses that many of the data points. The observed λ is then the random variable. Discuss the issue of bias in the context of these two distinct approaches. Can you work out a correction for the bias specifically for these cases? Beyond the issue of bias, are there other reasons to prefer one method over the other?

5.6 Stratification

Comparison of numerical efficiencies of computing μ_α^{ex} either directly or on the basis of thermodynamic integration leads us to the discussion of an important trick for these calculations: *stratification*. We focus here on a specific example, but the general advantage of stratification can be considered in a much less specific context: statistical uncertainties are mitigated by a nonstatistical subdivision of the problem, solution of the subdivided problems, and then recomposition of the whole (Hammersley and Handscomb, 1964, Section 5.3; Kalos and Whitlock, 1986, Section 4.5; Press *et al.*, 1992, Section 7.8).

Consider statistical evaluation of

$$e^{-\beta\mu_\alpha^{\text{ex}}} = \int \mathcal{P}_\alpha^{(0)}(\varepsilon)e^{-\beta\varepsilon}\,d\varepsilon \approx \frac{1}{m}\sum_{j=1}^{m}\left[e^{-\beta\Delta U_\alpha}\right]_j, \qquad (5.66)$$

as with Eq. (3.5), p. 33. To achieve a simple comparison we won't involve importance sampling which could be additionally advantageous. When m is large, the variance of this estimator is approximately

$$\sigma^2 = \frac{1}{m}\left[\left\langle\!\left\langle e^{-2\beta\Delta U_\alpha}\right\rangle\!\right\rangle_0 - \left\langle\!\left\langle e^{-\beta\Delta U_\alpha}\right\rangle\!\right\rangle_0^2\right]$$

$$= \frac{1}{m}e^{-\beta\mu_\alpha^{\text{ex}}}\left[\left\langle e^{-\beta\Delta U_\alpha}\right\rangle - \left\langle\!\left\langle e^{-\beta\Delta U_\alpha}\right\rangle\!\right\rangle_0\right]. \qquad (5.67)$$

The physical interpretation of the contributions in the brackets is that the system additionally prefers lower energy configurations as the interactions get turned on. Thus the first of the terms in brackets is larger than the second term, as it must be.

For hard-core interactions this variance is $p(1-p)/m$, with $p \leq 1$, familiar as the variance for the case of Bernoulli sampling that applies with hard-core insertions. How should the sample size be adjusted when the thermodynamic state, and hence p, is adjusted? The interesting circumstance is when p is small. Intuitively, we expect the sample size must be larger than $m \approx 1/p$ for credible results.

This point is also supported by the natural identification of σ/p as an indicator of the fractional statistical error in the partition function, or the error in the logarithm, i.e., in the desired free energy. It is also this ratio that contributes to the bias in the classic view of Eq. (5.63), p. 118. When $p \ll 1$, therefore, we expect that the sample size should be scaled as $m \sim (1-p)/p \sim e^{-\ln p}$. The qualitative conclusion suggested is that the sample size has to grow proportionally to the exponential of the entropy change that is sought.

Now consider an alternative calculation for a corresponding case of a hard-sphere solute. We start from

$$\beta\mu_{\mathrm{HC}}^{\mathrm{ex}} = -\ln\left\langle\!\left\langle e^{-\beta\Delta U_{\mathrm{HC}}} \right\rangle\!\right\rangle_0, \tag{5.68}$$

and imagine increasing the radius of the spherical solute by an amount $d\lambda$. The change in the solvation free energy is

$$d\beta\mu_{\mathrm{HC}}^{\mathrm{ex}}(\lambda) = -\frac{\left[\left\langle\!\left\langle e^{-\beta\Delta U_{\mathrm{HC}}}(\lambda+d\lambda)\right\rangle\!\right\rangle_0 - \left\langle\!\left\langle e^{-\beta\Delta U_{\mathrm{HC}}}(\lambda)\right\rangle\!\right\rangle_0\right]}{\left\langle\!\left\langle e^{-\beta\Delta U_{\mathrm{HC}}}(\lambda)\right\rangle\!\right\rangle_0}. \tag{5.69}$$

In the numerator the varied term excludes solvent from a thin shell of width $d\lambda$, and volume $4\pi\lambda^2 d\lambda$, that is not excluded from the unvaried term. This is exclusion from the *core* – the unvaried sphere – plus exclusion from the *shell*. If we use the exclusion from the core as an importance function, and use the denominator on the basis of our rule for averages this becomes

$$d\beta\mu_{\mathrm{HC}}^{\mathrm{ex}}(\lambda) = \langle dn \rangle_\lambda$$
$$\equiv 4\pi\lambda^2 \rho G(\lambda)\, d\lambda. \tag{5.70}$$

The last of these equations introduces the customary notation of $\rho G(\lambda)$, which is the number density of excluded solvent centers in this shell. The first of these equations indicates that the change in the solvation free energy is the expectation of the number of solvent centers in the shell. This is an infinitesimal quantity because the shell is infinitesimally thin.

Because the shell is thin, it will mostly be unoccupied. The probability that a single solvent center is located in the shell, $p_\lambda(1)$, is an infinitesimal quantity, and the probability that more than one solvent center is located in the shell is a quantity of higher infinitesimal order. Therefore, $\langle dn \rangle_\lambda = p_\lambda(1)$. Since there are only two probabilities to be considered here, $p_\lambda(1)$ and $1 - p_\lambda(1)$, a Bernoulli sampling model is appropriate, and as we have seen above the variance of such an estimator is $p_\lambda(1)(1 - p_\lambda(1))/m$, where m is the number of independent observations.

These arguments lead us to the conclusions that the integrated change in the solvation free energy is

$$\sum_j p_{\lambda_j}(1) \sim 4\pi \int_0^\lambda \rho G\left(\lambda'\right) \lambda'^2 \, d\lambda' = \beta\mu_{HC}^{ex}\left(\lambda\right); \qquad (5.71)$$

and the variance of this estimator is

$$\sigma^2 = \frac{1}{m}\sum_j p_{\lambda_j}(1)\left(1 - p_{\lambda_j}(1)\right)$$

$$\approx \frac{1}{m}\sum_j p_{\lambda_j}(1) \sim \frac{\beta\mu_{HC}^{ex}}{m}, \qquad (5.72)$$

because $p_\lambda(1) \ll 1$. Thus, to maintain a satisfactory accuracy when the thermodynamic state is changed, λ being fixed of course, we should increase $m \sim \beta\mu_{HC}^{ex}$, in sharp contrast to the estimate $m \sim \exp\left[\beta\mu_{HC}^{ex}\right]$ above. Stratification is the reason for this improvement, and it would be a decisive improvement in serious applications.

The λ integration achieves this stratification for the present example. The coupling parameter integration of Eq. (5.55), p. 117, similarly stratifies that application, and the variance found for the estimator, Eq. (5.59), p. 118, is analogous to that of the example of this section. The colloquial *windowing* methods exploit this stratification precisely, and a variety of histogram methods rely on this too. Umbrella sampling is fundamentally importance sampling as we emphasize in Section 5.1, p. 100. But it is occasionally discussed with multiple umbrellas (Chandler, 1987), and that mixes the two distinct ideas.

Exercise

5.32 Explain why the comparison of $\sigma e^{\beta\mu^{ex}}$ utilizing Eq. (5.67) with σ of Eq. (5.72) is the proper comparison here.

6

Statistical tentacles

This chapter discusses several statistical mechanical theories that are strongly positioned in the historical sweep of the theory of liquids. They are chosen for inclusion here on the basis of their potential for utility in analyzing simulation calculations, and their directness in connecting to the other fundamental topic discussed in this book, the potential distribution theorem. Therefore *tentacles* can be understood as tentacles of the potential distribution theorem. From the perspective of the preface discussion, the theories presented here might be useful for discovery of models such as those discussed in Chapter 4. These theories are a significant subset of those referred to in Chapter 1 as "... both difficult and strongly established ..." (Friedman and Dale, 1977), but the present chapter does not exhaust the interesting prior academic development of statistical mechanical theories of solutions. Sections 6.2 and 6.3 discuss alternative views of chemical potentials, namely those of density functional theory and fluctuation theory.

6.1 The MM and KS expansions

The Mayer–Montroll (Mayer and Montroll, 1941) and Kirkwood–Salsburg (Kirkwood and Salsburg, 1953) expansions are storied parts of basic statistical thermodynamics (Stell, 1985), but have been neglected for practical purposes because of a lack of recognition of how simple and simplifying they can be.

We introduce results with the specific example of a hard-core solute that was previously considered in Section 4.3. The hard-core results give perspective for a direct generalization to more realistic interactions.

Inclusion–exclusion and the MM expansion

Consider again $p_\alpha(0|\mathcal{R}^n) = \exp\left[-\beta\tilde{\mu}_\alpha^{\text{ex}}(\mathcal{R}^n)\right]$ for hard-core solutes as in Section 4.3. The most immediate guiding theory is the *inclusion–exclusion* development (Reiss *et al.*, 1959; Reiss, 1977; Riordan, 1978; van Kampen, 1992):

$$p\left(0|\mathcal{R}^n\right) = \sum_{k=0}^{\infty}(-1)^k\left\langle\binom{m}{k}|\mathcal{R}^n\right\rangle_0. \qquad (6.1)$$

Here the random variable m is the number of solvent centers within the observation volume. As examples: $\langle m|\mathcal{R}^n\rangle_0$ is the expected number of centers within the observation volume, and $\frac{1}{2}\langle m\left(m-1\right)|\mathcal{R}^n\rangle_0 = \langle\binom{m}{2}|\mathcal{R}^n\rangle_0$ is the number of *pairs* of centers included. These are standard combinatorial results (Riordan, 1978), as discussed with Fig. 6.1, and can be depicted in physical applications as shown there.

These results are obtained straightforwardly from the potential distribution theorem. We write (Kirkwood and Salsburg, 1953)

$$e^{-\beta\mu_\alpha^{ex}(\mathcal{R}^n)} = \left\langle e^{-\beta\Delta U_\alpha}|\mathcal{R}^n\right\rangle_0$$

$$= \left\langle\prod_{j=1}^{m}e^{-\beta\Delta U_\alpha(j)}|\mathcal{R}^n\right\rangle_0 = \left\langle\prod_{j=1}^{m}\left[1+f_\alpha\left(j\right)\right]|\mathcal{R}^n\right\rangle_0, \qquad (6.2)$$

then expand the product, and collect terms with a specified number of factors of $f_\alpha\left(j\right) \equiv e^{-\beta\Delta U_\alpha(j)} - 1$. There are $\binom{m}{k}$ terms with k factors of f_α so

$$e^{-\beta\mu_\alpha^{ex}(\mathcal{R}^n)} = 1 + \left\langle\binom{m}{1}f_\alpha\left(1\right)|\mathcal{R}^n\right\rangle_0$$

$$+ \left\langle\binom{m}{2}f_\alpha\left(1\right)f_\alpha\left(2\right)|\mathcal{R}^n\right\rangle_0$$

$$+ \left\langle\binom{m}{3}f_\alpha\left(1\right)f_\alpha\left(2\right)f_\alpha\left(3\right)|\mathcal{R}^n\right\rangle_0 + \cdots \qquad (6.3)$$

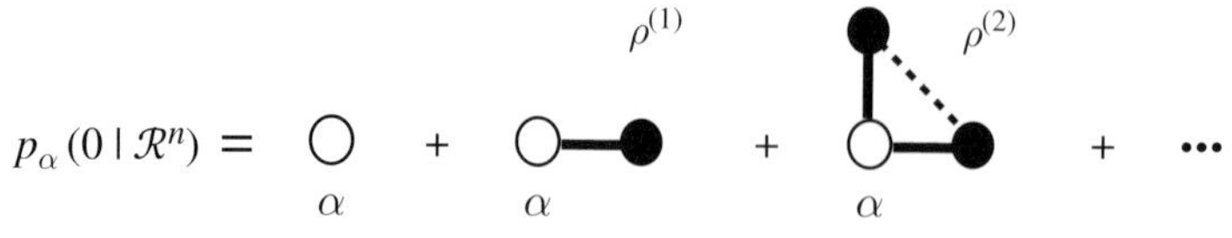

Figure 6.1 Mayer–Montroll expansion for the insertion probability $p(0|\mathcal{R}^n)$. The notation here is fairly standard; (see, for example, Hansen and McDonald, 1976; Andersen, 1977). The solid lines indicate factors of Mayer f functions introduced in Eq. (6.2) and are further discussed as Ursell functions beginning on p. 126. The inclusion–exclusion interpretation for hard-core cases is that the second term $-\langle m|\mathcal{R}^n\rangle_0$ assesses the m molecular volumes excluded to the distinguished solute. Then $\langle m\left(m-1\right)/2|\mathcal{R}^n\rangle_0$ corrects for pair overlaps of those excluded volumes. That pair correction generally needs a further triple overlap correction, and so on.

For the hard-core case, $f_\alpha(j) = -1$ when the jth molecule overlaps the observation volume for the distinguished species, and $f_\alpha(j) = 0$ otherwise. This is then the *inclusion–exclusion* Eq. (6.1) that motivated this discussion. Several important points can be made from these results. The first is that the primordial available volume model is obtained from the first two terms shown $\beta\tilde{\mu}_\alpha^{\text{ex}}(\mathcal{R}^n) \approx -\ln[1 - \langle m|\mathcal{R}^n\rangle_0]$. (See also the discussion of the Flory–Huggins model, Section 4.4, p. 78.) The second point is more basic: $p_\alpha(0|\mathcal{R}^n)$ is naturally expressed in terms of occupancy moments. The sum truncates sharply for cases where a finite maximum number of particles can be present in the observation volume. Note that the moments involved in the inclusion–exclusion series Eq. (6.1) are the same as the moments used in the information model Eq (4.32), p. 76. Since the latter procedure doesn't use this information in a series expansion, that information model procedure can be regarded as a resummation of the inclusion–exclusion series. In the general case for nonhard-core interactions, Eq. (6.3) with the general f functions involved is the Mayer–Montroll expansion.

Exercises

6.1 Evaluate the general term of the series Eq. (6.1) for the case that the particles are positioned randomly so that the Poisson distribution,

$$p_\alpha(k|\mathcal{R}^n) = \langle m|\mathcal{R}^n\rangle_0^{\,k}/k!\,\mathrm{e}^{-\langle m|\mathcal{R}^n\rangle_0},$$

applies. Confirm that this evaluation of Eq. (6.1) is consistent with the prediction that $p_\alpha(0|\mathcal{R}^n) = \mathrm{e}^{-\langle m|\mathcal{R}^n\rangle_0}$.

6.2 Show that if all terms of Eq. (6.1) with index $k > m$ are zero (due to the inability of fitting more than m particles in the observation volume), then the last nonzero term is the probability $p(m|\mathcal{R}^n)$.

6.3 For the case of a hard-core solute the condition $f \neq 0$ sharply defines the overlap volume. Generalize the combinatorial results above by considering the quantity

$$\left\langle \prod_{j=1}^{m} (1 + b_\alpha(j) + f_\alpha(j)) \,|\mathcal{R}^n \right\rangle_0 \qquad (6.4)$$

in which $f_\alpha(j) = -1$ if molecule j is in an observation volume and zero (0) otherwise, and $b_\alpha(j) \equiv -f_\alpha(j)$. Thus $b_\alpha(j)$ is one (1) in the observation volume and zero (0) otherwise. The average Eq. (6.4) is equal to one (1). If all $b_\alpha(j)$ were stricken, then the quantity Eq. (6.4) would be $p_\alpha(0|\mathcal{R}^n)$; see Eq. (6.1). By expanding and collecting terms ordered by the number of factors of $b(j)$ show that

$$k!\,p_\alpha(k|\mathcal{R}^n) = \sum_{j=0}^{} \frac{(-1)^j}{j!} \langle m(m-1)\ldots(m-k-j+1)|\mathcal{R}^n\rangle_0. \qquad (6.5)$$

Evaluate this average for the case of a Poisson distribution and thus confirm the correctness of this formula for that case.

Nonpair-decomposable interactions and Ursell functions

A difference between the study of simple models and the serious consideration of molecular liquids is the experimental possibility of interaction potential energies not representable by just a sum over molecular pairs. U is *pair-decomposable* if

$$U(\mathcal{N}) = \sum_{i=1}^{N} u^{(1)}(i) + \sum_{i>j=1}^{N} u^{(2)}(i, j). \tag{6.6}$$

The second sum is over $N(N-1)/2$ pairs of molecules. In the typical situation, the terms above would be examined according to the sequence

$$u^{(1)}(1) \equiv U(1) \tag{6.7}$$

$$u^{(2)}(1, 2) \equiv U(1, 2) - U(1) - U(2). \tag{6.8}$$

If a triplet contribution were required this would be represented as

$$U(\mathcal{N}) = \sum_{i=1}^{N} u^{(1)}(i) + \sum_{i>j=1}^{N} u^{(2)}(i, j) + \sum_{i>j>k=1}^{N} u^{(3)}(i, j, k), \tag{6.9}$$

and

$$u^{(3)}(1, 2, 3) \equiv U(1, 2, 3) - U(1, 2) - U(2, 3) - U(3, 1)$$
$$+ U(1) + U(2) + U(3). \tag{6.10}$$

The prescription for determining the functions $u^{(m)}$ is that an m-decomposable model should be correct if the system consisted only of m molecules. Formulae such as Eqs. (6.2) and (6.3) involve the Mayer f function that, for a pair decomposable case, is $f_\alpha(j) = \exp\left[-\beta u^{(2)}(0, j)\right] - 1$. A natural accommodation of nonpair-decomposable interactions in this case takes the goal of insuring that successive terms in a virial expansion are ordered by the density. This is the historical approach (Ursell, 1927), and is called an *Ursell expansion*. In this language, $f_\alpha(j)$ is an *Ursell function* (Stell, 1964; Münster, 1969). Again the idea is to require that the desired m-body Ursell function makes the product of Eq. (6.2) correct if just m molecules are involved. Thus for the case that only two molecules are involved

$$e^{-\beta \Delta U_\alpha} = e^{-\beta(U(0,1)-U(0)-U(1))} = 1 + f_\alpha^{(2)}(1),$$

$$f_\alpha^{(2)}(1) = e^{-\beta u^{(2)}(0,1)} - 1. \tag{6.11}$$

For the case that three molecules are involved we write

$$e^{-\beta \Delta U_\alpha} = e^{-\beta \left[u^{(2)}(0,1) + u^{(2)}(0,2) + u^{(3)}(0,1,2) \right]}$$

$$= 1 + f_\alpha^{(2)}(1) + f_\alpha^{(2)}(2) + f_\alpha^{(3)}(1) f_\alpha^{(3)}(2), \tag{6.12}$$

on the basis of the identification of Eq. (6.11), and the formal multiplication of the factors of $(1 + f_\alpha(j))$. ΔU_α here is the full potential energy of molecules (1,2,3) *less* the potential energy of the distinguished molecule (0), and the potential energy of the bath molecules (1,2) considered separately. Then

$$f_\alpha^{(3)}(1) f_\alpha^{(3)}(2) = e^{-\beta \left[u^{(2)}(0,1) + u^{(2)}(0,2) + u^{(3)}(0,1,2) \right]}$$

$$-e^{-\beta u^{(2)}(0,1)} - e^{-\beta u^{(2)}(0,2)} + 1$$

$$= f_\alpha^{(2)}(1) f_\alpha^{(2)}(2)$$

$$+ \left(1 + f_\alpha^{(2)}(1) \right) \left(1 + f_\alpha^{(2)}(2) \right) \left(e^{-\beta u^{(3)}(0,1,2)} - 1 \right). \tag{6.13}$$

It is just this product that is required in these expansions, so it isn't necessary to go further here. It should be clear how to write out the specific *j*th order term. We emphasize again that these approaches offer the possibility of considering series with only a finite number of non-negligible terms!

Exercise

6.4 Ursell decomposition can just as well be considered for representation of p-particle joint densities $\rho_{i_1 i_2 \ldots i_p}^{(p)}(r_1, \ldots, r_p)$ (Stell, 1964; Münster, 1969). For this more general problem, the idea is that when one subset of coordinates is distant from all others the distribution should gracefully adopt the form of independent distributions. For example, $\rho^{(2)}(r_1, r_2) \approx \rho^{(1)}(r_1) \rho^{(1)}(r_2)$ when r_1 and r_2 are well separated. Therefore, we would define Ursell functions

$$\rho^{(1)}(r_1) = \mathcal{U}^{(1)}(r_1)$$

$$\rho^{(2)}(r_1, r_2) = \mathcal{U}^{(2)}(r_1, r_2) + \mathcal{U}^{(1)}(r_1) \mathcal{U}^{(1)}(r_2), \tag{6.14}$$

so that

$$\mathcal{U}^{(2)}(r_1, r_2) = \rho^{(2)}(r_1, r_2) - \rho^{(1)}(r_1) \rho^{(1)}(r_2) \tag{6.15}$$

(A customary notation for fluids is that $\rho^{(2)}/\rho^2 \equiv g^{(2)}$ and $\mathcal{U}^{(2)}/\rho^2 \equiv h^{(2)} = g^{(2)} - 1$.) Consider how $\rho^{(3)}$ should be represented according to these ideas and work out $\mathcal{U}^{(3)}$ in terms of $\rho^{(p)}$ with $p \leq 3$.

Cumulant expansion

For convenience (Graham *et al.*, 1994), we use the *m*-to-the-*k*-falling notation $\langle m(m-1)\ldots(m-k+1)|\mathcal{R}^n\rangle \equiv \langle m^{\underline{k}}|\mathcal{R}^n\rangle$. Then Eq. (6.1) can be rewritten as

$$p_\alpha(0|\mathcal{R}^n) = \sum_{k=0}^\infty \frac{(-1)^k}{k!}\langle m^{\underline{k}}|\mathcal{R}^n\rangle_0 \equiv \langle e^{-m}|\mathcal{R}^n\rangle_0. \tag{6.16}$$

The last of these identifications is formal and symbolic; m^k must be interpreted as $m^{\underline{k}}$ in the expansion of this exponential.

Though this notation has some mnemonic value, it is helpful here in suggesting the naturalness of a cumulant expansion; the prominence of the $k!$ in Eq. (6.16) is the important issue, not the complication of the values for the moments. Further, the simplest approximation

$$\langle e^{-m}|\mathcal{R}^n\rangle_0 \approx e^{-\langle m|\mathcal{R}^n\rangle_0} \tag{6.17}$$

is already a sensible first step forward; this is the uncorrelated theory that would follow from the assumption of the Poisson distribution, as has been discussed above.

The formal manipulation of the moments that produces a cumulant expansion (see Eq. (5.18), p. 106) applies to the alternative moments considered here. Thus, for example,

$$\ln p(0|\mathcal{R}^n) \approx -\langle m|\mathcal{R}^n\rangle_0 + \frac{1}{2}\left[\langle m^2|\mathcal{R}^n\rangle_0 - \langle m|\mathcal{R}^n\rangle_0^2\right]$$

$$= -\langle m|\mathcal{R}^n\rangle_0 + \frac{1}{2}\left[\langle m(m-1)|\mathcal{R}^n\rangle_0 - \langle m|\mathcal{R}^n\rangle_0^2\right] \tag{6.18}$$

to 2nd order. Table 6.1 gives the combinations of moments through 4th order.

Exercise

6.5 Following Eq. (6.5) write

$$k!p(k|\mathcal{R}^n) = \sum_{j=0}^\infty \frac{(-1)^j}{j!}\langle m^{\underline{k+j}}|\mathcal{R}^n\rangle_0 \equiv \langle m^k e^{-m}|\mathcal{R}^n\rangle_0, \tag{6.19}$$

where the last identification must be interpreted symbolically to mean the formula in the middle of Eq. (6.19). In view of the Poisson distribution, the uncorrelated approximation should be

$$\langle m^k e^{-m}|\mathcal{R}^n\rangle_0 \approx \langle m|\mathcal{R}^n\rangle_0^k e^{-\langle m|\mathcal{R}^n\rangle_0}. \tag{6.20}$$

Table 6.1 *Successive contributions to the series*
$\beta \Delta \mu = \sum_1^{\infty} a_j / j!$ *of Figure 6.2. See also Table 5.1*

j	a_j
1	$+\langle m \rangle_0$
2	$-\langle m^2 \rangle_0 + \langle m \rangle_0{}^2$
3	$+\langle m^3 \rangle_0 - 3\langle m \rangle_0 \langle m^2 \rangle_0 + 2\langle m \rangle_0{}^3$
4	$-\langle m^4 \rangle_0 + 4\langle m \rangle_0 \langle m^3 \rangle_0 - 12\langle m \rangle_0{}^2 \langle m^2 \rangle_0 + 3\langle m^2 \rangle_0{}^2 + 6\langle m \rangle_0{}^4$

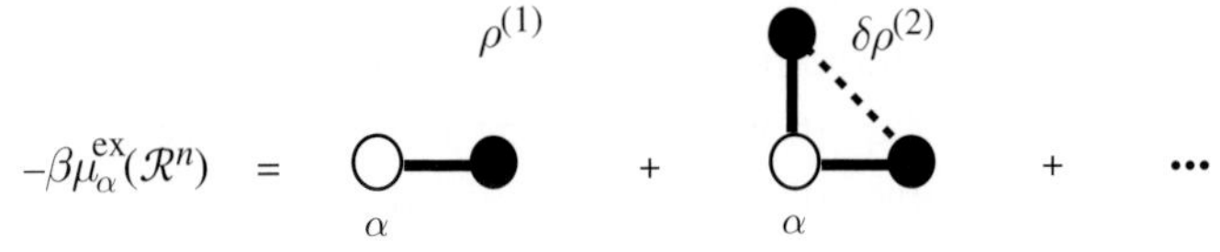

Figure 6.2 Cumulant expansion, notation as in Fig. 6.1. Here we assume full knowledge of the medium correlation functions in the absence of the solute; $\delta\rho^{(2)}(1,2) = \rho^{(2)}(1,2) - \rho^{(2)}(1)\rho^{(1)}(2)$. These contributions are ordered according to the number of bonds attached to the root point. Table 6.1 gives formulae for contributions through 4th order.

Note that

$$\langle m^k e^{-m} | \mathcal{R}^n \rangle_0 = \left[\frac{d^k}{d\lambda^k} \left\langle e^{\lambda m} | \mathcal{R}^n \right\rangle_0 \right]_{\lambda=-1} \qquad (6.21)$$

and devise a statistical approximation for $p_\alpha(k | \mathcal{R}^n)$ that is consistent with the 2nd-order result Eq. (6.18).

Kirkwood–Salsburg expansion

The idea for the potential distribution theorem, and consequently the Mayer–Montroll expansion Eq. (6.3), is to consider a distinguished, additional molecule. The idea for the Kirkwood–Salsburg expansion is to consider a distinguished, additional, pth atom conditional upon the locations of $(p-1)$ others. We will consider the $p = 2$ case specifically.

To this end, recall Eq. (3.50), p. 49:

$$g_{\alpha\gamma}^{(2)}(\boldsymbol{r}, \boldsymbol{r}') = e^{-\beta u_{\alpha\gamma}^{(2)}(\boldsymbol{r},\boldsymbol{r}')} \left[\frac{\left\langle e^{-\beta \Delta U_{\alpha\gamma}^{(2)}} | \boldsymbol{r}, \boldsymbol{r}' \right\rangle_0}{\left\langle e^{-\beta \Delta U_{\alpha}^{(1)}} | \boldsymbol{r} \right\rangle_0 \left\langle e^{-\beta \Delta U_{\gamma}^{(1)}} | \boldsymbol{r}' \right\rangle_0} \right]. \qquad (6.22)$$

We can re-express the factors in the denominator in the following way: one of those factors is used following Eq. (3.24), p. 42 to transform one of the distinguished test atoms into a real atom; the second of those factors is expressed with thermodynamic quantities on the basis of the potential distribution theorem to give

$$g_{\alpha\gamma}^{(2)}(\boldsymbol{r},\boldsymbol{r}') = \frac{z_\gamma}{\rho_\gamma \Lambda_\gamma^3} e^{-\beta u_{\alpha\gamma}^{(2)}(\boldsymbol{r},\boldsymbol{r}')} \left\langle e^{-\beta\left(\Delta U_{\alpha\gamma}^{(2)} - \Delta U_{\alpha}^{(1)}\right)} \middle| \boldsymbol{r},\boldsymbol{r}' \right\rangle_0 \tag{6.23}$$

or

$$g_{\alpha\gamma}^{(2)}(\boldsymbol{r},\boldsymbol{r}') = e^{-\beta u_{\alpha\gamma}^{(2)}(\boldsymbol{r},\boldsymbol{r}')} \left\langle e^{-\beta\left(\Delta U_{\gamma}^{(1)} - \mu_{\gamma}^{ex}\right)} \middle| \boldsymbol{r},\boldsymbol{r}' \right\rangle_0, \tag{6.24}$$

making the additive assumption that $\Delta U_{\alpha\gamma}^{(2)} = \Delta U_{\alpha}^{(1)} + \Delta U_{\gamma}^{(1)}$. If the deficit from pair decomposability of interactions is important, then the simplified notation in the last step to arrive at Eq. (6.24) should be borne in mind. Furthermore, the notation doesn't specifically emphasize that the atom of type α is actual and the atom of type γ is a test particle, so be careful. This hints at the fact that this formulation is not trivially symmetric when the two labeled particles are of the same type. But the quantity on the left is symmetric, so nice approximations should conform to this observation. Eq. (6.24) should be compared with Eq. (3.35), p. 45.

We now proceed in much the same way that was effective for Eq. (6.2):

$$e^{\beta\left(u_{\alpha\gamma}^{(2)}(\boldsymbol{r},\boldsymbol{r}') - \mu_{\gamma}^{ex}\right)} g_{\alpha\gamma}^{(2)}(\boldsymbol{r},\boldsymbol{r}') = 1$$

$$+ \left\langle \binom{m-1}{1} f_{\alpha\gamma}^{(2)}(2) \middle| \boldsymbol{r},\boldsymbol{r}' \right\rangle_0$$

$$+ \left\langle \binom{m-1}{2} f_{\alpha\gamma}^{(2)}(2) f_{\alpha\gamma}^{(2)}(3) \middle| \boldsymbol{r},\boldsymbol{r}' \right\rangle_0$$

$$+ \left\langle \binom{m-1}{3} f_{\alpha\gamma}^{(2)}(2) f_{\alpha\gamma}^{(2)}(3) f_{\alpha\gamma}^{(2)}(4) \middle| \boldsymbol{r},\boldsymbol{r}' \right\rangle_0 + \cdots$$

$$\tag{6.25}$$

A depiction of these results is shown in Fig. 6.3. Notice that the Mayer–Montroll expansion for $e^{-\beta\mu_{\gamma}^{ex}}$ is obtained for large separations where both $\exp\left[-\beta u_{\alpha\gamma}^{(2)}(\boldsymbol{r},\boldsymbol{r}')\right]$ and $g_{\alpha\gamma}^{(2)}(\boldsymbol{r},\boldsymbol{r}')$ are practically one (1). The terms in the MM expansion of Eq. (6.3) can be matched up in a one-to-one fashion with terms in the KS expansion Eq. (6.25). In this way, approximate use of this series for $g_{\alpha\gamma}^{(2)}(\boldsymbol{r},\boldsymbol{r}')$ can correspond to an approximate result for $e^{-\beta\mu_{\gamma}^{ex}}$ and vice versa. Furthermore, a term-by-term subtraction introduces the Ursell function $h_{\alpha\gamma}^{(2)} = g_{\alpha\gamma}^{(2)} - 1$ on the left, an initial $f_{\alpha\gamma}^{(2)}$ on the right, and simplifies the integrals to be considered on the right.

Aside from potential utility as generators of approximate physical theories for $g_{\alpha\gamma}^{(2)}(\boldsymbol{r},\boldsymbol{r}')$, these Kirkwood–Salsburg expansions permit systematic construction of

$$e^{\beta(u-\mu^{ex})} g^{(2)} =$$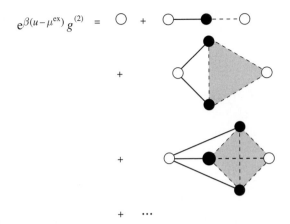

Figure 6.3 Compare with Fig. 6.1. Here again the solid lines indicate factors of Mayer f functions as in Eq. (6.25) and further discussed as Ursell functions beginning on p. 126. The second term shown is $\int f(r')\rho^{(1)}(r'\,|r)\mathrm{d}^3r'$ for a simple fluid. The shaded regions with $m-1$ black disks and one white disk represent conditional densities $\rho^{(m-1)}(r_1,\ldots,r_{n-1}\,|r)$. That white disk, rightmost here, corresponds to the *real* atom positioned at r on which the averages are conditioned. The other white disk, leftmost here, corresponds to the *test* particle.

series for μ_α^{ex} that are ordered in powers of the density, known as *virial expansions*. The way this can work is that the Kirkwood–Salsburg equations permit elimination of distribution functions in the Mayer–Montroll series (Fig. 6.1), through the necessary order in the density.

Exercises

6.6 Consider the KS expansion, Fig. 6.3, applied to a one-component fluid and truncated at the second term displayed; this will be satisfactory for low density because the subsequent terms have higher-power initial density multipliers. What is the corresponding MM approximate theory for the excess chemical potential? Show that this KS approximate theory, expressed for $h^{(2)}$, is

$$e^{-\beta\mu^{ex}} h^{(2)}(r) = f(r)\left(1 + \int f(r)\rho\mathrm{d}^3r\right)$$

$$+ (1+f(r))\int f(r')h^{(2)}(r'-r)\rho\mathrm{d}^3r'. \qquad (6.26)$$

Consider a hard-core system for which $h^{(2)} = -1 = f$ inside the core. What can you conclude about the consistency of this simple theory with the approximate result for the excess chemical potential?

6.7 Consider using Eq. (6.25), p. 130, to determine the form of $g^{(2)}(r, r')$ as a density expansion. Use that approach together with the cumulant expansion (Fig. 6.2) to find μ^{ex} that is correct to $O(\rho^2)$.

6.8 Consider the hard-sphere fluid and establish the zero separation theorem $\left[e^{\beta u^{(2)}(r,r')} g^{(2)}(r, r') \right]_{r=r'} = e^{\beta \mu^{\text{ex}}}$ for that case. Explain how Eq. (6.25), p. 130, achieves this result.

6.9 Consider a primitive model of a dilute electrolyte solution: the system is composed of ions of two types that interact as

$$u_{\eta\nu}(|r - r'|) = \frac{q_\eta q_\nu}{\epsilon |r - r'|}. \tag{6.27}$$

ϵ is interpreted as the dielectric constant of the solvent. The system is at high temperature (small β), and low densities conforming to bulk solution neutrality $\sum \rho_\eta q_\eta = 0$. Short-ranged repulsive interactions between ions prevent collapse but won't be present in the final form here. Consider the approximation to $g_{\eta\nu}^{(2)}$ that corresponds to the first line of Fig. 6.3. Explain why $f_{\eta\nu}^{(2)} \approx -\beta u_{\eta\nu}^{(0)}$ and $\mu_\eta^{\text{ex}} \approx 0$ in the present case. With comparable approximations, obtain the Debye–Hückel integral equation:

$$h_{\eta\nu}^{(2)}(|r - r'|) \approx -\beta u_{\eta\nu}^{(0)}(|r - r'|)$$

$$- \int \beta u_{\eta\gamma}^{(0)}(|r - r''|) \rho_\gamma h_{\gamma\nu}^{(2)}(|r'' - r'|) \mathrm{d}^3 r'' \tag{6.28}$$

with $h_{\eta\nu}^{(2)}(|r - r'|) = g_{\eta\nu}^{(2)}(|r - r'|) - 1$.

6.2 Density functional and classic integral equation theories

Our goal in this section is to establish a common density functional language for considering solution thermodynamics, and then to provide some density functional perspective on the chemical potentials that are the principal actors in this book. Those results are the foremost goal of this section. But we then use those results to follow Percus's development (Percus, 1964) of classic integral equations for the structure of liquids, and to make further observations that provide more perspective on these issues.

Functionals and functional derivatives

A rule that assigns to each function in a suitable set a number in another set is a *functional* (Mathews and Walker, 1964; Stakgold, 1979). According to some practices, a basic view of the theory of solutions is to find, study, and characterize the functionals which yield the thermodynamic properties when the intermolecular interactions are supplied. Thus, some of the tools of functional analysis are commonplace in our subject, and identifications of some functional derivatives are basic. Here we catalog the necessary functional derivative relations.

The functional derivative relations we consider here are defined as follows (Hansen and McDonald, 1976, see Section 4.2; Rowlinson and Widom, 1982, see Section 4.5). Consider a functional $\mathcal{A}$ of functions $f(x)$; this will be denoted by $\mathcal{A}[f(x)]$. Then consider a small change $f \to f + \delta f$. An evaluation of the change

$$\mathcal{A}[f + \delta f] = \mathcal{A}[f] + \int \frac{\delta \mathcal{A}}{\delta f(x)} \delta f(x) \mathrm{d}x \qquad (6.29)$$

to linear order in δf identifies the indicated kernel

$$\frac{\delta \mathcal{A}}{\delta f(x)} = \frac{\delta \mathcal{A}}{\delta f(x)}[f] \qquad (6.30)$$

as the functional derivative. The simplest example is obtained by considering the form of the contribution of the internal energy of a simple atomic system from a generic external field $\varphi(\boldsymbol{r})$:

$$\mathcal{U}[\varphi(\boldsymbol{r})] = \int_{\mathcal{V}} \varphi(\boldsymbol{r}) \rho(\boldsymbol{r}) \mathrm{d}^3 r. \qquad (6.31)$$

Then

$$\frac{\delta \mathcal{U}}{\delta \varphi(\boldsymbol{r})} = \rho(\boldsymbol{r}). \qquad (6.32)$$

A small handful of functional derivatives underlie the discussions that will follow. Our initial step will be to consider the Helmholtz free energy $\mathcal{A}$, introduced in Eq. (2.15), p. 26, and how it depends on an external potential energy field of the form anticipated by Eq. (6.31). In the absence of such an external field, the Helmholtz free energy is a function of $(\boldsymbol{n}, \mathcal{V}, T)$, $\mathcal{A} = \mathcal{A}(\boldsymbol{n}, \mathcal{V}, T)$. Thus we now generalize our questions to consider $\mathcal{A}[\varphi(\boldsymbol{r})] \equiv \mathcal{A}(\boldsymbol{n}, \mathcal{V}, T)[\varphi(\boldsymbol{r})]$. Our evaluation will be based upon the classical expression

$$\mathcal{A}[\varphi(\boldsymbol{r}) + \delta\varphi(\boldsymbol{r})] - \mathcal{A}[\varphi(\boldsymbol{r})] =$$

$$- kT \ln \left\langle \exp\left[-\beta \sum_{\alpha, j} \delta\varphi_\alpha(\boldsymbol{r}_j)\right] \right\rangle. \qquad (6.33)$$

(Compare this with Eq. (2.27), p. 28.) The average on the right is the thermal average for the interacting system corresponding to the conditions $(\boldsymbol{n}, \mathcal{V}, T)[\varphi(\boldsymbol{r})]$. (If there are particles of several types, then we consider fields $\varphi_\alpha(\boldsymbol{r})$ for each type; the right side of Eq. (6.33) makes that clear, but we will not have an explicit

notation like that for the left side.) It is then an algebraic exercise, one that we
have encountered already in Sections. 4.1 and 5.3, to show that

$$A[\varphi(r) + \delta\varphi(r)] - A[\varphi(r)] = \int_{\mathcal{V}} \langle \rho_\alpha(r) \rangle \, \delta\varphi_\alpha(r) \mathrm{d}^3 r$$

$$- \frac{\beta}{2} \int_{\mathcal{V}} \int_{\mathcal{V}} \langle \delta\rho_\alpha(r) \delta\rho_\gamma(r') \rangle \, \delta\varphi_\alpha(r) \delta\varphi_\gamma(r') \mathrm{d}^3 r \mathrm{d}^3 r' + \cdots \quad (6.34)$$

Thus,

$$\frac{\delta A}{\delta\varphi_\alpha(r)} [\varphi(r)] = \langle \rho_\alpha(r) \rangle, \quad (6.35)$$

and the second derivative is

$$\frac{\delta\rho_\alpha(r)}{\delta\varphi_\gamma(r')} = -\beta \langle \delta\rho_\alpha(r) \delta\rho_\gamma(r') \rangle \equiv \chi_{\alpha\gamma}(r, r'). \quad (6.36)$$

The latter coefficient addresses the question of the response of the system density
to an applied external field, and is called the *susceptibility*. The final functional
derivative relation that we will use views this susceptibility – Eq. (6.36) – as
a matrix, considers the matrix inverse, and is a generalization of the partial
derivative relation

$$\delta_{\alpha\gamma} = \sum_{\eta} \left(\frac{\partial f_\alpha}{\partial x_\eta} \right)\bigg|_{x_\nu, \nu \neq \eta} \left(\frac{\partial x_\eta}{\partial f_\gamma} \right)\bigg|_{f_\zeta, \zeta \neq \gamma}. \quad (6.37)$$

Then the analogous relation

$$\delta_{\alpha\gamma} \delta(r - r') = \int_{\mathcal{V}} \sum_{\eta} \frac{\delta\rho_\alpha(r)}{\delta\varphi_\eta(r'')} \frac{\delta\varphi_\eta(r'')}{\delta\rho_\gamma(r')} \mathrm{d}^3 r'' \quad (6.38)$$

permits an evaluation of the inverse susceptibility

$$\frac{\delta\varphi_\alpha(r)}{\delta\rho_\gamma(r')} \equiv \chi_{\alpha\gamma}^{-1}(r, r'). \quad (6.39)$$

Pushing the analogy of Eq. (6.37) further leads to the suggestion that χ^{-1} might
be naturally a functional of ρ rather than φ. In the context of the theory of liquids,
this inverse susceptibility is written as

$$-\beta \frac{\delta\varphi_\alpha(r)}{\delta\rho_\gamma(r')} = \frac{\delta_{\alpha\gamma} \delta(r - r')}{\rho_\alpha(r)} - c_{\alpha\gamma}(r, r'), \quad (6.40)$$

which introduces the Ornstein–Zernike (OZ) direct correlation function $c_{\alpha\gamma}(r, r')$.
Note again that

$$\langle \delta\rho_\alpha(r) \delta\rho_\gamma(r') \rangle = \rho_\alpha(r) \delta_{\alpha\gamma} \delta(r - r') + \rho_\alpha(r) \rho_\gamma(r') h_{\alpha\gamma}(r, r'); \quad (6.41)$$

for context, see Eq. (6.15), p. 127. Here we drop the superscript (2), and consider this Ursell function $h_{\alpha\gamma}$ of particles of species type α and γ. We can consider again the inverse relation Eq. (6.38), which becomes

$$h_{\alpha\gamma}(r,r') = c_{\alpha\gamma}(r,r') + \int_{\mathcal{V}} c_{\alpha\eta}(r,r'')\rho_\eta(r'')h_{\eta\gamma}(r'',r')\, d^3r''.$$

(6.42)

This is the Ornstein–Zernike (OZ) equation from which the function $c_{\alpha\gamma}(r,r')$ acquires its name.

The functional Eq. (6.33) suggests that it is most natural to consider the Helmholtz free energy as a functional of the external potential energy field: $\mathcal{A} = \mathcal{A}[\varphi_\alpha(r)]$. Our interest will be in density functionals. To pursue this we introduce the Legendre transform (Callen, 1985), $\mathcal{A} - \sum_\alpha \int_{\mathcal{V}} \varphi_\alpha(r)\rho_\alpha(r)\, d^3r \equiv W[\rho_\alpha(r)]$. Consider a change $\delta\varphi_\alpha(r)$ in the external potential. Of course, the equilibrium density will change by $\delta\rho_\alpha(r)$, and further

$$\delta W = \int_{\mathcal{V}} \left(\frac{\delta\mathcal{A}}{\delta\varphi_\alpha(r)} - \rho_\alpha(r) \right) \delta\varphi_\alpha(r)\, d^3r$$

$$- \int_{\mathcal{V}} \varphi_\alpha(r)\, \delta\rho_\alpha(r)\, d^3r.$$

(6.43)

In view of Eq. (6.35), if we use the equilibrium density corresponding to $\varphi_\alpha(r)$, then the first of these terms vanishes, and

$$\varphi_\alpha(r) = -\frac{\delta W}{\delta\rho_\alpha(r)},$$

(6.44)

a functional of the density, $\rho_\alpha(r)$. If this functional were satisfactorily known, we might consider solving for the equilibrium density given the field. The second functional derivative is then

$$\frac{\delta^2\beta W}{\delta\rho_\alpha(r)\,\delta\rho_\gamma(r')} = \frac{\delta_{\alpha\gamma}\delta(r-r')}{\rho_\alpha(r)} - c_{\alpha\gamma}(r,r').$$

(6.45)

Density functional perspective on chemical potentials

Our target for these formalities is a density functional perspective on chemical potentials. Notice that for a given $\varphi_\alpha(r)$ both $\mathcal{A}[\varphi]$ (Eq. (6.33)) and $\rho_\alpha(r)$ (Eq. (6.35)) are determined. We might consider an alteration of the density pattern, $\delta\rho_\alpha(r)$, but without a changed external field. The altered density is not the equilibrium result of a change of the external field, but is just a hypothetical different density. For this virtual variation in the density, the free energy change is

$$\delta\mathcal{A} = \int_{\mathcal{V}} \sum_\alpha \left(\varphi_\alpha(r) + \frac{\delta W}{\delta\rho_\alpha(r)} \right) \delta\rho_\alpha(r) d^3r.$$

(6.46)

This comes from $\mathcal{A} \equiv \sum_\alpha \int_V \varphi_\alpha(\mathbf{r})\rho_\alpha(\mathbf{r})\mathrm{d}^3 r + \mathcal{W}[\rho_\alpha(\mathbf{r})]$ as above, but with the virtual density variation not requiring the equilibrium conditions implicit in Eq. (6.44). From here we proceed with the usual Lagrange multiplier calculation to make this free energy stationary for variations that don't change the particle number:

$$\delta(A - \sum_\alpha \mu_\alpha n_\alpha) = \int_V \sum_\alpha \left(\varphi_\alpha(\mathbf{r}) + \frac{\delta \mathcal{W}}{\delta \rho_\alpha(\mathbf{r})} - \mu_\alpha \right) \delta \rho_\alpha(\mathbf{r}) \mathrm{d}^3 r, \qquad (6.47)$$

where μ_α are the necessary Lagrange multipliers. Requiring the free energy to be stationary with respect to now unrestricted variations in the density implies the condition of equilibrium:

$$\varphi_\alpha(\mathbf{r}) - \mu_\alpha = -\frac{\delta \mathcal{W}}{\delta \rho_\alpha(\mathbf{r})}. \qquad (6.48)$$

This differs from Eq. (6.44) by a constant. But in hindsight we recognize that the derivation of Eq. (6.44), following from Eq. (6.33), was blind to changes in the numbers of molecules, and thus permitted a spatially uniform constant to settle the question of net changes in molecule numbers.

The thermodynamic interpretation of the spatially uniform constant μ_α is found by considering the case that the solution of Eq. (6.48) is found. Using that solution in the formula Eq. (6.46) gives

$$\delta A = \sum_\alpha \mu_\alpha \delta n_\alpha. \qquad (6.49)$$

Thus μ_α is the thermodynamic chemical potential.

The first contribution of Eq. (6.45) is easy to integrate. So we can write

$$\frac{\delta^2 \beta \mathcal{W}^{\mathrm{ex}}}{\delta \rho_\alpha(\mathbf{r}) \delta \rho_\gamma(\mathbf{r}')} = -c_{\alpha\gamma}(\mathbf{r}, \mathbf{r}'), \qquad (6.50)$$

and

$$\beta \mu_\alpha = \ln \Lambda_\alpha^3 + \ln \rho_\alpha(\mathbf{r}) + \beta \varphi_\alpha(\mathbf{r}) + \frac{\delta \beta \mathcal{W}^{\mathrm{ex}}}{\delta \rho_\alpha(\mathbf{r})}$$

$$= \ln \rho_\alpha(\mathbf{r}) \Lambda_\alpha^3 + \beta \varphi_\alpha(\mathbf{r}) + \frac{\delta \beta \mathcal{W}^{\mathrm{ex}}}{\delta \rho_\alpha(\mathbf{r})}, \qquad (6.51)$$

where $\ln \Lambda_\alpha^3$ is the required constant of integration; this identification is justified by the known result for the noninteracting atomic case, Eq. (2.4), p. 24. Notice especially the superscript ^{ex} in the notation $\mathcal{W}^{\mathrm{ex}}$. This Eq. (6.51) should be compared with Eq. (3.35), p. 45:

$$\frac{\delta \beta \mathcal{W}^{\mathrm{ex}}}{\delta \rho_\alpha(\mathbf{r})} = -\ln \left\langle e^{-\beta \Delta U_\alpha} | \mathbf{r} \right\rangle_0. \qquad (6.52)$$

With these formulae the inverse question is formally simple. Given the density, and the functional we could determine the external field from Eq. (6.51)

$$\beta\varphi_\alpha(r) = -\ln\rho_\alpha(r) - \frac{\delta\beta W^{ex}}{\delta\rho_\alpha(r)}, \qquad (6.53)$$

to within a constant.

These results establish that

$$c_{\alpha\gamma}(r, r') = -\frac{\delta\beta\mu_\alpha^{ex}(r)}{\delta\rho_\gamma(r')}[\rho], \qquad (6.54)$$

and that the excess chemical potential obtained by integrating this information is

$$\beta\mu_\alpha^{ex}(r)[\rho] = \frac{\beta\delta W^{ex}}{\delta\rho_\alpha(r)}. \qquad (6.55)$$

Nonuniformities: PY and HNC integral equations

These results permit a useful expression for the interaction contribution to the chemical potential when the system is nonuniform:

$$\exp\{-\beta\mu_\alpha^{ex}(r)[\rho]\}$$
$$= \exp\{-\beta\mu_\alpha^{ex}(r)[\bar\rho] + \sum_\gamma \int_V c_{\alpha\gamma}(r, r')[\rho_\gamma(r') - \bar\rho_\gamma)]d^3r' + \cdots\}. \qquad (6.56)$$

Here $\bar\rho_\gamma$ are uniform densities providing the basis for a functional Taylor series. Then the subsequent terms are ordered in the differences $[\rho_\gamma(r') - \bar\rho_\gamma]$ and the coefficients such as $c_{\alpha\gamma}(r, r')$ – see Eq. (6.54) – are properties of the uniform system at densities $\bar\rho_\gamma$.

The relation Eq. (6.56) is key to Percus's derivation of the Percus–Yevick approximation (Percus, 1964). We consider the case where the inhomogeneity is produced by the location of a distinguished particle of type ν at the origin, and inquire about the surrounding fluid. The Kirkwood–Salsburg formula Eq. (6.24), p. 130, offers the interpretation of the left side of Eq. (6.56):

$$\exp\{-\beta\mu_\alpha^{ex}(r)[\rho]\} = \frac{\bar\rho_\alpha g_{\alpha\nu}(r)}{z_\alpha e^{-\beta u_{\nu\alpha}(r)}}. \qquad (6.57)$$

To obtain the Percus–Yevick approximation, we neglect second-order and higher contributions, and the right side of Eq. (6.56) is evaluated as

$$\exp\left\{-\beta\mu_\alpha^{ex}(;r)[\bar\rho] + \sum_\gamma \int_V c_{\alpha\gamma}(r, r')[\rho_\gamma(r') - \bar\rho_\gamma)]d^3r'\right\}$$
$$= \frac{\bar\rho_\alpha}{z_\alpha}\exp\left\{\sum_\gamma \int_V c_{\alpha\gamma}(r, r')[\rho_\gamma(r') - \bar\rho_\gamma)]d^3r'\right\}$$

$$\approx \frac{\bar{\rho}_\alpha}{z_\alpha}\left\{1+\sum_\gamma \int_V c_{\alpha\gamma}(\boldsymbol{r},\boldsymbol{r}')\left[\rho_\gamma(\boldsymbol{r}')-\bar{\rho}_\gamma)\right]\mathrm{d}^3r'\right\}$$

$$=\frac{\bar{\rho}_\alpha}{z_\alpha}\left\{1+\sum_\gamma \rho_\gamma\int_V c_{\alpha\gamma}(\boldsymbol{r},\boldsymbol{r}')h_{\gamma\nu}(\boldsymbol{r}')\mathrm{d}^3r'\right\}. \tag{6.58}$$

Combining, using the OZ equation, and cleaning up, we obtain

$$e^{\beta u_{\nu\alpha}(r)}g_{\alpha\nu}(r)=1+h_{\alpha\nu}(r)-c_{\alpha\nu}(r), \tag{6.59}$$

or

$$c_{\alpha\nu}(r)=(1-e^{\beta u_{\nu\alpha}(r)})g_{\alpha\nu}(r)=f_{\alpha\nu}(r)e^{\beta u_{\nu\alpha}(r)}g_{\alpha\nu}(r), \tag{6.60}$$

the PY approximation. Notice that if the range of intermolecular interactions is finite, as is the case for hard-core interactions, then $c_{\alpha\nu}(r)$ is zero outside that range. On the other hand, Eq. (6.59) shows that for hard-core interactions $g_{\alpha\nu}(r)$ will be zero for overlapped configurations in this approximation, and that is physically correct.

The hypernetted chain approximation results if we don't linearize, so

$$e^{\beta u_{\nu\alpha}(r)}g_{\alpha\nu}(r)=\exp\{h_{\alpha\nu}(r)-c_{\alpha\nu}(r)\}, \tag{6.61}$$

or

$$g_{\alpha\nu}(r)=\exp\{-\beta u_{\nu\alpha}(r)+h_{\alpha\nu}(r)-c_{\alpha\nu}(r)\}, \tag{6.62}$$

the HNC approximation.

Further perspective: the HLR theory

The *hydrostatic linear-response* (HLR) theory (Chen and Weeks, 2003) gives further perspective on the PY theory, and exercises the preceding concepts of this section. We can illustrate that development starting with the questions of the density distortion of a solution in the neighborhood of a distinguished molecule. As with the Kirkwood–Salsburg approach, the distinguished molecule is viewed as a source of an external field exerted on the solution. The linear-response perspective on that density distortion is

$$\rho(\boldsymbol{r})\left[\varphi\right]\approx\rho(\boldsymbol{r})\left[\bar{\varphi}\right]+\int_V \chi(\boldsymbol{r},\boldsymbol{r}')\left[\bar{\varphi}\right]\left(\varphi(\boldsymbol{r}')-\bar{\varphi}\right)\mathrm{d}^3r', \tag{6.63}$$

where $\bar{\varphi}$ is a uniform field that produces the density $\bar{\rho}$, i.e. $\rho(\boldsymbol{r})\left[\bar{\varphi}\right]=\bar{\rho}\left[\bar{\varphi}\right]$. $\chi(\boldsymbol{r},\boldsymbol{r}')\left[\bar{\varphi}\right]$ is the susceptibility of the liquid experiencing the uniform field $\bar{\varphi}$. It is a standard idea (Pratt *et al.*, 1988; 1990; Hoffman and Pratt, 1990; 1991) to try to optimize this description by considering each r in turn, finding a $\bar{\varphi} = \bar{\varphi}_r$ that does a good job for each r. In particular, the solution of the nonlinear equation

$$\bar{\varphi}_r = \frac{\int_V \chi(r, r')\,[\bar{\varphi}_r]\,\varphi(r')\mathrm{d}^3 r'}{\int_V \chi(r, r')\,[\bar{\varphi}_r]\,\mathrm{d}^3 r'} \tag{6.64}$$

annuls the linear-response contribution, and provides a field to use in the locally uniform approximation $\rho(r)\,[\varphi] \approx \bar{\rho}\,[\bar{\varphi}_r]$.

It is interesting also to ask the inverse question: what is the field corresponding to an observed density? Then

$$\varphi(r)\,[\rho] \approx \varphi(r)\,[\bar{\rho}] + \int_V \chi^{-1}(r, r')\,[\bar{\rho}]\,(\rho(r') - \bar{\rho})\,\mathrm{d}^3 r'. \tag{6.65}$$

The field on the right side, $\varphi(r)\,[\bar{\rho}]$, is the one that produces a uniform density $\bar{\rho}$; we could represent that uniform case entirely with a chemical potential as

$$\varphi(r)\,[\rho] \approx \mu\,(\bar{\rho}) + \int_V \chi^{-1}(r, r')\,[\bar{\rho}]\,(\rho(r') - \bar{\rho})\,\mathrm{d}^3 r'. \tag{6.66}$$

The optimization condition here is

$$0 = \int_V \chi^{-1}(r, r')\,[\bar{\rho}]\,(\rho(r') - \bar{\rho}_r)\,\mathrm{d}^3 r', \tag{6.67}$$

or, in view of Eq. (6.40), p. 134

$$\rho(r) = \bar{\rho}_r + \bar{\rho}_r \int_V c(r, r')\,[\bar{\rho}_r]\,(\rho(r') - \bar{\rho}_r)\,\mathrm{d}^3 r', \tag{6.68}$$

which is suggestively similar to the OZ equation, Eq. (6.42), p. 135.

Equation (6.68) might be used in two different ways. The first way is to answer the question originally posed: given $\rho(r)$, what is $\varphi(r)$? The density $\rho(r)$, the uniform system chemical potential $\mu\,(\bar{\rho})$, and the direct correlation function for the uniform solution must be known for this purpose.

Another use is available, however, if $\varphi(r)$ is known. Then we can view Eq. (6.67) as an insistence on the *hydrostatic* choice for $\bar{\rho}_r$; that is with $\mu\,(\bar{\rho})$ known, choose $\mu\,(\bar{\rho}_r) = \varphi(r)$, assuming that a solution for $\bar{\rho}_r$ exists on the basis of this requirement. Finally, if $c(r, r')\,[\bar{\rho}_r]$ is available too, Eq. (6.68) might be solved for the full density response $\rho(r)$. The interesting point of perspective is that for the case of a hard-core solute, Eq. (6.68) implies the PY theory because the hydrostatic choice sets $\bar{\rho}_r = 0$, and thus $\rho(r) = 0$, for overlapping configurations; but for non-overlap configurations $\bar{\rho}_r = \rho$, the bulk density. Then, Eq. (6.68) is

$$\rho(r) - \rho = \rho \int_V c(r, r')\,[\bar{\rho}_r]\,(\rho(r') - \rho)\,\mathrm{d}^3 r' \tag{6.69}$$

for all nonoverlapping configurations. But this corresponds to the usual OZ equation without an explicit contribution from the OZ direct correlation, that is with the assumption that such a contribution is zero for nonoverlapping configurations, as in the PY approximation for hard-core interactions.

Further perspective: HNC and RISM theories of molecular liquids

A beautiful, strongly established feature of the classic theory of liquids, but too involved for our current discussion, is the formal determination (Morita and Hiroike, 1961, specifically Eq. 4.22; Stell, 1964, specifically Eqs. 10–12) of the interaction contributions to the chemical potential on the basis of system densities and observed molecule–molecule correlations for the case of pair decomposable intermolecular interactions. Those developments are dazzlingly complete – for the case of pair interactions – but formal in the sense that they involve an infinite series with unproven convergence in specific cases of interest.

With the HNC approximation this development is explicit and concise. For cases where the HNC approximation is known to be usefully accurate, it would be interesting to attempt to evaluate those developments more fully on the basis of correlation functions observed in simulations, in order to learn something about the characteristics of successive terms in those series in a favorable case.

This point of interest is brought forward by the RISM approach to the structure of molecular liquids, and a RISM model with HNC closure supports a similar result for the excess chemical potential in terms of atom–atom correlations (Singer and Chandler, 1985; Hirata, 1998). "RISM" – reference interaction site model – is an acronym that refers to a class of theories for the joint two-atom distributions in molecular liquids. The most basic decision of RISM models is that theories of molecular liquids should focus first on the atom–atom distributions extracted from X-ray and neutron scattering data rather than more complex possibilities; this highly practical point was not so obvious in an earlier epoch when models of molecular liquids were scarcely realistic on an atomic scale. That basic decision was encapsulated by invention of a site–site (or atom–atom) Ornstein–Zernike (SSOZ) (Cummings and Stell, 1982) equation that involved <u>intra</u>molecular atom–atom correlations. The original suggestions (Chandler and Andersen, 1972) were sufficiently successful as to support subsequent flamboyant developments, and to be substantially impervious to more fundamental improvements (Chandler *et al.*, 1982). For these reasons a full discussion of the RISM models wouldn't fit here. Fortunately, a devoted exposition of current RISM work is already available (Hirata, 1998).

6.3 Kirkwood–Buff theory

The Kirkwood–Buff theory of solutions (Kirkwook and Buff, 1951) doesn't depend on special assumptions about the nature of the intermolecular interactions such as pairwise additivity, and can be based upon the relation Eq. (6.54) which leads to the density derivatives of the interaction contributions to chemical potentials, and in this way leads to evaluations of those chemical potentials. The KB approach has been of significant practical importance (Mansoori and Matteoli, 1990).

We obtain the KB relations by considering the change in the interaction contribution to the chemical potential of species α upon increasing the density of species γ with temperature fixed:

$$\delta\mu_\alpha^{ex} = \int_V \frac{\delta\mu_\alpha^{ex}}{\delta\rho_\gamma(r)} \delta\rho_\gamma(r) d^3r. \tag{6.70}$$

We consider a fluid, and take the density change to be uniform in the system volume $\delta\rho_\gamma(r) = \Delta\rho_\gamma$. Noting also Eq. (6.54), this leads to

$$\left(\frac{\beta\partial\mu_\alpha^{ex}}{\partial\rho_\gamma}\right)\Bigg|_{\rho_\nu, \nu\neq\gamma} = -\int_V c_{\alpha\gamma}(r) d^3r. \tag{6.71}$$

This relation can be exploited by obtaining information on the correlations as described by Eq. (6.42) and using the OZ relation to obtain the integral of the direct correlation function. A density integration would then typically permit reconstruction of μ_α^{ex} relative to a defined standard state. We have already encountered the interesting point that the density derivative on the left of Eq. (6.71) is ill-defined for the case of ionic interactions in the limit of low concentrations; the interaction contributions to chemical potentials of ions vary as $\sqrt{\rho}$ for vanishing concentrations, see p. 93. This leads to the conclusion that the integral on the right of Eq. (6.71) is ill-defined in that situation.

Exercise

6.10 Consider the case of a spherical solute (A) dissolved at infinite dilution in a molecular solvent (S), and show that the partial molar volume – see Eq. (4.100), p. 97 – can be expressed as

$$v_A = -kT\kappa_T - 4\pi \int_0^\infty (g_{AS}(r) - 1) r^2 dr, \tag{6.72}$$

where $g_{AS}(r)$ is the radial distribution of solvent centers from the solute, and κ_T is the isothermal compressibility of the solvent. Hint: see Pratt and Pohorille (1993); Paulaitis *et al.* (1994). Give an interpretation of the factor of κ_T. Give a qualitative drawing of $g_{AS}(r)$, and discuss consequences of the features of that drawing for v_A.

7

Quasi-chemical theory

An initial discussion of a quasi-chemical approach appeared in Section 4.6. This chapter gives a fuller development of those ideas. The idea of our initial discussion was to introduce a statistical model capable of a natural description of strong association phenomena in solutions, and the example of ion clustering in electrolyte solutions was considered. But the quasi-chemical approach may be founded on broader concepts, and given a more extensive development. The most primitive idea is to identify an inner-shell region from the rest of the neighborhood of a distinguished solute, and to rely on a painstaking treatment of the inner shell, with full molecular resolution. The remainder of the neighborhood of that distinguished solute – the outer-shell region – can be given an alternative statistical description, and then a proper matching of results for inner and outer shells must be accomplished. The pragmatic approach of using alternative methods for physically distinct spatial regions is important.

Many problems of solution theory cry out for chemical treatment of an obvious inner shell. For example, complexes such as $Fe(H_2O)_6{}^{3+}$ naturally present themselves as important solution structures when $Fe^{3+}(aq)$ is considered. But discussion of the thermodynamics of $Fe^{3+}(aq)$ on that basis requires a satisfactory parsing of the thermochemistry associated with the ligand species. In evaluating the standard Gibbs free energy of a complex solution species, what contribution should be assigned to the ligands? Should those contributions be a standard Gibbs free energy for those ligand species? Or should those ligand contributions correspond to the actual activity of the ligands? The quasi-chemical theory supports the latter alternative: the absolute activity – including the ideal concentration dependence – is directly involved in constructing the standard free energy of $Fe^{3+}(aq)$.

The appellation *quasi-chemical* is acquired from Guggenheim (Guggenheim, 1935; 1938). This is because central equations of the theory have a structure that is familiar from chemical considerations; this is true of the developments below too. The independent and simultaneous theory of Bethe (1935), soon shown to be essentially equivalent (Fowler and Guggenheim, 1940), was called the "method of local configurations" by Kirkwood (1940), and that idea is a characteristic of the present considerations as well.

Despite the kinship of the present development to these historical works, we don't try to follow upon the historical footsteps directly. Part of our reservation is that these historical works were based upon lattice-gas models that are archaic as contributions to molecular science: sometimes useful, but rarely an attempt to come to grips with molecular problems at a molecular resolution. Nevertheless, quasi-chemical and Bethe theories are probably the *most effective* approximate physical theories available for those lattice gas problems, and an important goal of our present discussion will be to discover corresponding theories of molecular science, as distinct from lattice gases.

It is interesting that Eq. (4.98), p. 97, offers a discrete-state partition function for the description of the inner-sphere contribution to the thermodynamics. But the discrete coordinate is an occupation number for a precisely defined configurational region, and parameters required for this discrete-state partition function are obtained by molecular-level calculations. Therefore, molecular realism isn't the first casualty of these theories, although strong approximations are typically accumulated after the formulation of quasi-chemical theories.

7.1 Derivation of the basic quasi-chemical formula

This quasi-chemical development can be motivated by the observation that many solute–solvent interactions can be characterized as chemical associations – strong relative to the thermal energy, kT – although not necessarily covalent interactions. The interactions considered here are typically both short-ranged and structurally specific. So we can identify an inner shell or proximal region around the solute that will accommodate strongly associating solvent molecules, and a second region, the outer shell, that corresponds to the remainder of the system volume. For protein solutions, we might define this inner shell to include regions the size of water molecules in close proximity to solvent-accessible, hydrophilic groups that are likely to bind water molecules. The definition of a hydrogen bond between pairs of water molecules, or between a water molecule and a particular amino acid on the surface of a protein are customary considerations in describing hydrogen-bonding interactions in aqueous solutions. The inner shell can be specified by defining an indicator function $b_\alpha (j)$, with $b_\alpha (j) = 1$ when solvent molecule j occupies the inner shell of a distinguished molecule of type α, and $b_\alpha (j) = 0$ when this solvent molecule is outside that region. With this definition, a natural outer-shell contribution to the potential distribution theorem expression for the solute excess chemical potential is

$$\beta \mu_\alpha^{ex}\big|_{\text{outer}} = -\ln \left\langle\!\!\left\langle e^{-\beta \Delta U_\alpha} \prod_j [1 - b_\alpha(j)] \right\rangle\!\!\right\rangle_0 . \qquad (7.1)$$

Note that the additional factor within the average, the $\prod_j (1 - b_\alpha(j))$, would be zero for any solvent configuration in which a solvent molecule is found in the inner shell. Thus, this expression involves a potential distribution average under the constraint that no binding in the inner shell is permitted. We can formally write the full expression for the excess chemical potential as

$$\beta \mu_\alpha^{ex} = -\ln \left[\frac{\left\langle\!\!\left\langle e^{-\beta \Delta U_\alpha} \right\rangle\!\!\right\rangle_0}{\left\langle\!\!\left\langle e^{-\beta \Delta U_\alpha} \prod_j [1 - b_\alpha(j)] \right\rangle\!\!\right\rangle_0} \right]$$

$$- \ln \left\langle\!\!\left\langle e^{-\beta \Delta U_\alpha} \prod_j [1 - b_\alpha(j)] \right\rangle\!\!\right\rangle_0 . \qquad (7.2)$$

This expression separates solute–solvent interactions into an outer-shell contribution, for which simpler physical approximations might be helpful, and an inner-shell contribution, which we consider from the perspective of quasi-chemical theory.

Note that the ratio of averages appearing in Eq. (7.2) is the form considered in Section 3.3, p. 41. The averaged quantity is just the indicator function $\prod_j [1 - b_\alpha(j)]$. The ratio evaluates the average with the solute present. Since this function is equal to one (1) for precisely those cases that the defined inner shell is empty, and is equal to zero (0) otherwise, we can recast Eq. (7.2) as

$$\beta \mu_\alpha^{ex} = \ln x_0 - \ln \left\langle\!\!\left\langle e^{-\beta \Delta U_\alpha} \prod_j [1 - b_\alpha(j)] \right\rangle\!\!\right\rangle_0 , \qquad (7.3)$$

where x_0 is the probability of those configurations for which the solute has no solvent molecules bound in the inner shell, i.e., it is the ordinary average of the indicator function:

$$x_0 = \frac{\left\langle\!\!\left\langle e^{-\beta \Delta U_\alpha} \prod_j [1 - b_\alpha(j)] \right\rangle\!\!\right\rangle_0}{\left\langle\!\!\left\langle e^{-\beta \Delta U_\alpha} \right\rangle\!\!\right\rangle_0}$$

$$= \left\langle \prod_j [1 - b_\alpha(j)] \right\rangle . \qquad (7.4)$$

Chemical equilibrium description of x_0

The conclusive step in this quasi-chemical development is to recognize that a model for stoichiometric chemical equilibrium provides a correct description of x_0. We imagine following a specific solute molecule of interest through chemical conversions defined as changes in the inner-shell populations,

$$\alpha + n\text{W} \rightleftharpoons \alpha\text{W}_n, \tag{7.5}$$

where α indicates the solute – for example, a protein molecule – and W a solvent species – water is the most common one. When these transformations are in equilibrium,

$$\frac{c_n}{c_0} = K_n\rho_{\text{W}}{}^n \tag{7.6}$$

with c_n indicating the concentration of an αW_n complex. Equation (7.6) serves as a definition of K_n and is just the usual *products over reactants* chemical equilibrium ratio.

Supplying the appropriate normalization to evaluate x_0 explicitly gives

$$x_0 \equiv \frac{c_0}{\sum_n c_n} = \frac{1}{1 + \sum\limits_{n \geq 1} K_n\rho_{\text{W}}{}^n} \tag{7.7}$$

and substituting into Eq. (7.3) gives the desired final result,

$$\beta\mu_\alpha^{\text{ex}} = -\ln\left[1 + \sum_{n \geq 1} K_n\rho_{\text{W}}{}^n\right] - \ln\left\langle\!\!\left\langle e^{-\beta\Delta U_\alpha}\prod_j[1 - b_\alpha(j)]\right\rangle\!\!\right\rangle_0. \tag{7.8}$$

Equation (7.8) offers a clear separation of inner-shell and outer-shell contributions so that different physical approximations might be used in these different regions, and then matched. The description of inner-shell interactions will depend on access to the equilibrium constants K_n. These are well defined, observationally and computationally (see Eq. (7.10)), and so might be the subject of either experiments or statistical thermodynamic computations. For simple solutes, such as the Li^+ ion, *ab initio* calculations can be carried out to obtain approximately the K_n (Pratt and Rempe, 1999; Rempe *et al.*, 2000; Rempe and Pratt, 2001), on the basis of Eq. (2.8), p. 25. With definite quantitative values for these coefficients, the inner-shell contribution in Eq. (7.8) appears just as a function involving the composition of the defined inner shell. We note that the net result of dividing the excess chemical potential in Eq. (7.8) into inner-shell and outer-shell contributions should not depend on the specifics of that division. This requirement can provide a variational check that the accumulated approximations are well matched.

Exercise

7.1 Show that the traditional chemical thermodynamic consideration of a chemical transformation such as

$$n_A A + n_B B \rightleftharpoons n_C C + n_D D \tag{7.9}$$

with the potential distribution theorem Eq. (3.18), p. 40, leads to

$$
K \equiv \frac{\rho_C{}^{n_C} \rho_D{}^{n_D}}{\rho_A{}^{n_A} \rho_B{}^{n_B}}
$$

$$
= \frac{\left(\langle\!\langle e^{-\beta \Delta U_C}\rangle\!\rangle_0 \frac{q_C^{\text{int}}}{\Lambda_C{}^3}\right)^{n_C} \left(\langle\!\langle e^{-\beta \Delta U_D}\rangle\!\rangle_0 \frac{q_D^{\text{int}}}{\Lambda_D{}^3}\right)^{n_D}}{\left(\langle\!\langle e^{-\beta \Delta U_A}\rangle\!\rangle_0 \frac{q_A^{\text{int}}}{\Lambda_A{}^3}\right)^{n_A} \left(\langle\!\langle e^{-\beta \Delta U_B}\rangle\!\rangle_0 \frac{q_B^{\text{int}}}{\Lambda_B{}^3}\right)^{n_B}}, \tag{7.10}
$$

generalizing Eq. (2.8), p. 25. Writing this as

$$
K = K^{(0)} \frac{\left(\langle\!\langle e^{-\beta \Delta U_C}\rangle\!\rangle_0\right)^{n_C} \left(\langle\!\langle e^{-\beta \Delta U_D}\rangle\!\rangle_0\right)^{n_D}}{\left(\langle\!\langle e^{-\beta \Delta U_A}\rangle\!\rangle_0\right)^{n_A} \left(\langle\!\langle e^{-\beta \Delta U_B}\rangle\!\rangle_0\right)^{n_B}}, \tag{7.11}
$$

displays the ideal $K^{(0)}$ explicitly. The conclusion is that the equilibrium ratios are indeed well defined as a matter of observation and of molecular computation.

7.2 Clustering in more detail

The preceding general derivation didn't seek details that can serve to ground the theory better. One ingredient for deriving the quasi-chemical pattern is a counting device used to sort configurations according to proximity. If ΔU_α were pairwise decomposable we would write

$$
e^{-\beta \Delta U_\alpha} = \prod_j [1 + f_\alpha(j)], \tag{7.12}
$$

where, as in Section 6.1, p. 123, $f_\alpha(j)$ is the Mayer f (cluster) function describing the interactions between the α solute and the solvent molecule j. Series expansions then proceed in a direct and simple way for the simple case to which this equation applies. The clustering features of these expansions are valuable, and we can preserve them in cases that don't present pairwise decomposable ΔU_α by writing

$$
1 = \prod_j [1 + b_\alpha(j) + f_\alpha(j)]. \tag{7.13}
$$

$b_\alpha(j)$ is one (1) in a geometrically defined αj-bonding region and zero (0) otherwise; $b_\alpha(j)$ is an indicator function (van Kampen, 1992). $f_\alpha(j)$ is then defined by

$$
f_\alpha(j) = -b_\alpha(j). \tag{7.14}
$$

With this set-up the $f_\alpha(j)$ is analogous to a Mayer f function for a hard object and can play the same role of monitoring the description of packing effects in liquids. See Exercise 6.1, p. 125 for an example.

The strategy for our derivation will be to insert this *resolution of unity*, Eq. (7.13), within the averaging brackets of the potential distribution theorem, then expand and order the contributions according to the number of factors of $b_\alpha(j)$ that appear. We emphasize that physical interactions are not addressed here and that the hard-core interactions associated with discontinuity in $f_\alpha(j)$ appear for counting purposes only.

Low-order contributions

Let's note some of the properties of this expansion, and of the terms that result. Consider initially the term that is zeroth order, with no factors of $b_\alpha(j)$ appearing. That term is

$$0th \ order \ in \ b_\alpha(j): \prod_j [1 + f_\alpha(j)].$$

This would be the interaction potential energy for the bath with a hard object that excludes the bath from any bonding region, because then $f_\alpha(j) = -1$ and the statistical weight would be zero.

$$1st \ order \ in \ b_\alpha(j): Nb_\alpha(1) \prod_{j \neq 1} [1 + f_\alpha(j)].$$

There is just one bath or solvent molecule in the bonding region and N possibilities for that specific molecule due to the existence of N solvent molecules in the system. All other molecules are excluded from the bonding region.

$$2nd \ order \ in \ b_\alpha(j): \left(\frac{N(N-1)}{2}\right) b_\alpha(1) b_\alpha(2) \prod_{j \notin \{1,2\}} [1 + f_\alpha(j)].$$

Now there are two bath or solvent molecules in the bonding region, that pair could have been chosen in $N(N-1)/2$ ways, and all pairs other than the specific pair are excluded.

General term

The pattern of the contributions to this cluster expansion is obvious. If we take the general formula for the mth-order expansion term and now include the Boltzmann factor for the full system of solute and solvent molecules, then we can illustrate the equivalence of two different views of the total system. In one view the system is divided into a distinguished solute and N solvent molecules. In an alternative view the distinguished solute is a complex cluster. In the cluster view, the solute is surrounded by m solvent *ligand* molecules in the bonding region with the remaining $N - m$ solvent molecules occupying the external region. The full mechanical potential energy is parsed as

$$U(\alpha) + U(\mathcal{N}) + \Delta U_\alpha = U(\alpha + m) + U(\mathcal{N} - m) + \Delta U_{\alpha + m}. \tag{7.15}$$

None of the energies considered here need be pairwise decomposable. The general mth term then adopts the form

$$e^{-\beta[U(\alpha+m)+U(\mathcal{N}-m)+\Delta U_{\alpha+m}]} \binom{N}{m} \left(\prod_{i \in m} b_\alpha(i) \right) \left(\prod_{j \notin m} [1 + f_\alpha(j)] \right)$$

$$= \binom{N}{m} e^{-\beta U(\mathcal{N}-m)} e^{-\beta U(\alpha+m)} \left(\prod_{i \in m} b_\alpha(i) \right) e^{-\beta \Delta U_{\alpha+m}} \left(\prod_{j \notin m} [1 + f_\alpha(j)] \right). \tag{7.16}$$

The factor $\binom{N}{m}$ is the number of ways of selecting m specific solvent molecules as occupants of the inner shell; those are selected for the m factors of $b_\alpha(i)$.

The remaining factors of Eq. (7.16) will make up the integrand involved in a canonical partition function. But we will prefer to evaluate a grand canonical partition function. In doing this, we will select the appropriate m factors of the activity out of the activities available, and denote that combination by z^m. The $\binom{N}{m}$ feature will be zero for cases that N isn't sufficient to supply the ligand set m. Therefore, all nonzero contributions will be proportional to the factor of z^m.

There will be activities $z^{(N-m)}$ left aside from such a contribution to the desired grand canonical partition function. The factor $\binom{N}{m}$ supplies a numerator of $N!$ that cancels exactly that factor in the denominator of the original grand partition function population weight; $\binom{N}{m}$ also contributes a handy denominator of the form $(N - m)!$ and leaves another denominator factor of $m!$. This argument amounts to

$$\frac{z^N}{N!} e^{-\beta[U(\alpha+m)+U(\mathcal{N}-m)+\Delta U_{\alpha+m}]} \binom{N}{m} \left(\prod_{i \in m} b_\alpha(i) \right) \left(\prod_{j \notin m} [1 + f_\alpha(j)] \right)$$

$$= \frac{z^{(N-m)}}{(N-m)!} e^{-\beta U(\mathcal{N}-m)} \frac{z^m}{m!} e^{-\beta U(\alpha+m)} \left(\prod_{i \in m} b_\alpha(i) \right) e^{-\beta \Delta U_{\alpha+m}} \left(\prod_{j \notin m} [1 + f_\alpha(j)] \right). \tag{7.17}$$

Then, as with Eq. (3.17), p. 40, we multiply and divide by $\mathcal{Q}(N - m)$ to compose the sought-after grand canonical population weight. Finally, we multiply and divide to factor-forward a configurational partition function associated with the weight

$$e^{-\beta U(\alpha+m)} \prod_{i\in m} b_\alpha(i) \tag{7.18}$$

for the $\alpha+m$ complex. The canonical partition function for that defined case will be denoted by $\mathcal{Q}(\alpha+m)/m!$. Then

$$\text{mth term contributing to } \left\langle\!\!\left\langle e^{-\beta\Delta U_\alpha}\right\rangle\!\!\right\rangle_0:$$

$$\left\langle\!\!\left\langle e^{-\beta\Delta U_{\alpha+m}} \prod_{j\notin m} [1+f_\alpha(j)]\right\rangle\!\!\right\rangle_0 \mathcal{Q}(\alpha+m)\frac{z^m}{m!}.$$

Using the notation $\mathcal{Q}(\alpha+m)/m! = V q^{\text{int}}_{\alpha+m}/\Lambda^3_{\alpha+m}$ and collecting these results, finally we get the intended partition function formula

$$e^{-\beta\mu^{\text{ex}}_\alpha} = \sum_{m\geq 0} \frac{q^{\text{int}}_{\alpha+m}\Lambda^3_\alpha}{q^{\text{int}}_\alpha\Lambda^3_{\alpha+m}} \left\langle\!\!\left\langle e^{-\beta\Delta U_{\alpha+m}} \prod_{j\notin m} [1+f_\alpha(j)]\right\rangle\!\!\right\rangle_0 z^m. \tag{7.19}$$

Compare this result with Eq. (4.93), p. 96. After noting again that $f_\alpha(j) = -b_\alpha(j)$, the $m = 0$ term is recognized as the outer-shell contribution of Eq. (7.1) p. 144.

Exercise

7.2 Use the general potential distribution theorem to eliminate the explicit factors z^m from Eq. (7.19) to obtain Eq. (7.8), p. 145.

7.3 An example of a primitive quasi-chemical treatment: Be²⁺(aq)

To fix the ideas in a simple case, let's discuss an example of a primitive quasi-chemical treatment of the statistical thermodynamics of Be²⁺(aq). This closely follows Asthagiri and Pratt (2003).

Beryllium metal has properties that make it technologically attractive (Alderighi *et al.*, 2000), but these advantages are counterbalanced by the toxicity of inhaled beryllium dust, which can cause chronic beryllium disease (Sauer *et al.*, 2002). The etiology of this immune hyper-response disease (Fontenot *et al.*, 2001) is poorly understood, but the final disease state is characterized by lung failure.

At low concentrations and neutral pH conditions, Be²⁺(aq) can hydrolyze water and then form colloidal aggregates linked with HO⁻ species (Alderighi *et al.*, 2000; Bruno, 1987), as suggested by Fig. 7.1. These are chemical issues.

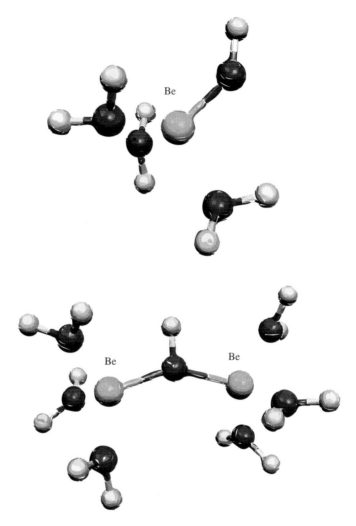

Figure 7.1 Structures representing the deprotonated tetra-aquo cation, $BeOH\,(H_2O)_3^+$, and the aggregate formed by the coalescence of one such unit and a $Be\,(H_2O)_4^{2+}$ ion, with expulsion of a water molecule.

The inner-shell combinations pertinent to a quasi-chemical treatment of Be^{2+} hydration are

$$Be^{2+} + nH_2O \rightleftharpoons Be(H_2O)_n^{2+}, \tag{7.20}$$

and the equilibrium ratios K_n are defined for these chemical processes. A simple quasi-chemical approximation for the hydration free energy of such ions is

$$\beta\mu_{Be^{2+}}^{ex} \approx -\ln\left[K_{\bar{n}}^{(0)}\rho_{H_2O}^{\bar{n}}\right] + \beta\mu_{Be(H_2O)_{\bar{n}}^{2+}}^{ex} - \bar{n}\beta\mu_{H_2O}^{ex}. \tag{7.21}$$

$K_n^{(0)}$ are those coefficients evaluated for the case that the effects of a medium external to the n-ligand cluster are neglected. Those parameters can be obtained by calculating free energy changes for these reactions with standard programs to treat electronic structure.

The largest term of the sum of Eq. (7.8), p. 145, is indicated by $\bar{n}$ in Eq. (7.21). Limiting that sum to the largest term makes a negligible difference in these situations. There are two reasons for that. The first is that succeeding terms in such a sum reflect chemical energy differences. A slight advantage in energy on a chemical scale for one term means that it dominates all others on a thermal energy scale. For example, if two terms in the general sum were precisely the same, the present neglect would entail an error in the free energy of $kT \ln 2$, which is a negligible magnitude here. The second reason is that there are only a finite, and small, number of terms in the full sum, so there isn't a chance to accumulate a significant entropic contribution in that way. A more significant entropic contribution is the temperature dependence of the density displayed in Eq. (7.21).

The density ρ_{H_2O} appearing in Eq. (7.21) is the actual density of liquid water. At standard conditions this is roughly $1\,\mathrm{g\,cm^{-3}}$. Since $K_n^{(0)}$ is calculated for an ideal gas, with standard state pressure of 1 atm, the effect of the actual density is a *replacement contribution* of $-nkT \ln(\rho_{H_2O}/\rho_0) \approx -nkT \ln(1354)$, where ρ_0 is the density of water at the standard state pressure of 1 atm and temperature of 298 K. This amounts to a contribution of $-4.27\,\mathrm{kcal\,mol^{-1}}$ per ligand. This contribution is favorable to this reaction, and reflects the higher availability of ligands at the density of the liquid, 1354 times higher than that of the standard state.

The parameters $K_n^{(0)}$ evaluated in this way express no influence of the medium external to the cluster. Consulting Eq. (7.11), p. 146, we see that we should incorporate that influence through contributions such as

$$- \ln K_n = - \ln K_n^{(0)} + \beta \mu_{\mathrm{Be(H_2O)}_n^{2+}}^{\mathrm{ex}} - \beta \mu_{\mathrm{Be(H_2O)}_{n=0}^{2+}}^{\mathrm{ex}} - n \beta \mu_{\mathrm{H_2O}}^{\mathrm{ex}}. \qquad (7.22)$$

But the contribution $\beta \mu_{\mathrm{Be(H_2O)}_{n=0}^{2+}}^{\mathrm{ex}}$ is precisely the outer-shell contribution obtained from Eq. (7.8), p. 145. Thus this contribution is cancelled exactly by the outer-shell contribution that is acknowledged just at this point. This is how the approximation Eq. (7.21) is obtained.

A dielectric model, as discussed in fundamental form in Section 4.2, p. 67, was adopted for the results of Fig. 7.2. It is found that $\bar{n} = 4$, not surprisingly. The calculated hydration free energy of $-567.7\,\mathrm{kcal\,mol^{-1}}$ is in reasonable agreement with the value of $-574.6\,\mathrm{kcal\,mol^{-1}}$ cited in Marcus (1985).

In contrast to more typical applications of a dielectric model, the final result here is not very sensitive to the radius parameters that are used. There are two reasons for that insensitivity. First, the radius parameters used for ligand molecules

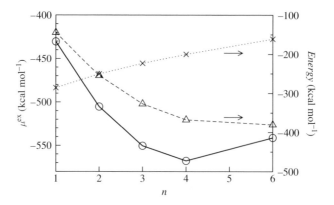

Figure 7.2 Quasi-chemical contributions of the hydration free energy of $Be^{2+}(aq)$. Cluster geometries were optimized using the B3LYP hybrid density functional (Becke, 1993) and the 6-31+G(d, p) basis set. Frequency calculations confirmed a true minimum, and the zero point energies were computed at the same level of theory. Single-point energies were calculated using the 6-311+G(2d, p) basis set. A purely inner-shell $n = 5$ cluster was not found; the optimization gave structures with four (4) inner-sphere water molecules and one (1) outer-sphere water molecule. For $n = 6$ both a purely inner-shell configuration, and a structure with four (4) inner-shell and two (2) outer-shell water molecules were obtained. The quasi-chemical theory here utilizes only the inner-shell structure. $\bigcirc: -kT \ln \left[K_n^{(0)} \rho_{H_2O}^n \right] - n\mu_{H_2O}^{ex} + \mu_{Be(H_2O)_n^{2+}}^{ex}$ (left ordinate) vs. n. $\triangle:$ $-kT \ln K_n^{(0)} - nkT \ln(1354)$; $\times: \mu_{Be(H_2O)_n^{2+}}^{ex} - n\mu_{H_2O}^{ex}$. The inner shell was defined by a Be–O radius of 2.0 Å. Using a smaller radius did not make an appreciable difference. Note that the outer-shell contribution is about half of the inner-shell, so neither contribution is irrelevant. For further details, see Asthagiri and Pratt (2003).

are the same for the difference of Eq. (7.21), and the variations expected in each term track each other to some extent. Second, the exact cancellation of the outer-shell contribution means that a radius parameter used for the species at the core of the complex, the beryllium ion, is largely irrelevant because the ligands serve effectively to bury that species. Of course, exact cancellation is another case of a difference between two terms with variations in one term compensating for the variations of the other. Variational qualities will be discussed further in the following section.

The acidity of $Be(H_2O)_4^{2+}$ is described by

$$K_a = \frac{\left[BeOH(H_2O)_3^+ \right] \left[H^+ \right]}{\left[Be(H_2O)_4^{2+} \right]}, \tag{7.23}$$

corresponding to the reaction

$$\mathrm{Be(H_2O)_4^{2+} \rightleftharpoons BeOH(H_2O)_3^+ + H^+} \tag{7.24}$$

under standard conditions. A detailed discussion of this extension of our present quasi-chemical activity would take us too far afield for now. We can note, however, that natural further calculations yield $pK_a \approx 3.8$. By fitting experimental free energy changes for the case of low total Be^{2+} concentration (Alderighi *et al.*, 2000; Bruno, 1987), it is found that $\mathrm{Be(H_2O)_4^{2+}}$ exists in appreciable amounts only below a pH of 3.5. The present calculated pK_a is in good agreement with these observations. This value of pK_a has the standard interpretation that the deprotonated complex $\mathrm{BeOH(H_2O)_3^+}$ is a thousand times more probable than $\mathrm{Be(H_2O)_4^{2+}}$ at neutral pH.

7.4 Analysis of *ab initio* molecular dynamics

This section continues the discussion of the quasi-chemical theory, but incorporates AIMD (*ab initio* molecular dynamics) calculations to develop the point that the central quantities of quasi-chemical theories can be obtained from simulation calculations. Additionally, we develop a variational perspective on the quasi-chemical partitioning of inner and outer shells. The calculations which provide the basis of this discussion treat liquid water (Asthagiri *et al.*, 2003c).

We discuss the latter topic – the variational character – first. The inner-shell and outer-shell terms displayed in Eq. (7.8) offer the possibility of balancing contributions. In fact, the sum of those two terms would be independent of the definition of the inner shell if no further approximations were made. This is because the rearrangement induced by definition of the inner shell was entirely formal and correct, and the original problem was independent of the definition of the inner shell.

When approximations are made in the evaluation of those distinct terms, the sum needn't be independent of definition of the inner shell. We argue here that when the sum is insensitive to local adjustment of the inner shell then the inevitable approximations are well balanced. The quasi-chemical approach is variational in this sense.

We now develop an example of this variational character. We utilize results from *ab initio* molecular dynamics (AIMD) for that purpose, and estimate $\mu_{\mathrm{H_2O}}^{\mathrm{ex}}$ for liquid water. The *ab initio* molecular dynamics simulations were carried out with the VASP (Kresse and Hafner, 1993; Kresse and Furthmüller, 1996) simulation program, as described in detail in Asthagiri *et al.* (2003c). *Ab initio* molecular dynamics of aqueous solutions are recent activities compared with other simulation methods for aqueous solutions, and basic characterization of the new methods is still underway; see Grossman *et al.* (2004) and Schwegler *et al.* (2004) for initial examples.

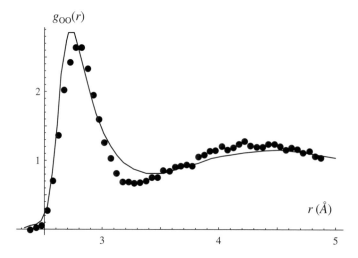

Figure 7.3 Oxygen–oxygen radial density distribution for liquid water. The smooth curve is the experimental result of Head-Gordon and Hura (2002) the dots are the data, collected in bins of width 0.05 Å, from *ab initio* molecular dynamics utilizing the rPBE electron density functional. The estimated temperature is 314 K for the latter case, slightly higher than that of the experiments at 300 K. See Asthagiri *et al.* (2003c).

Figure 7.3 shows the OO radial distribution function. Note that the first minimum of $g_{OO}(r)$ is around 3.3 Å, and this suggests an inner-shell radius for a quasi-chemical analysis.

We focus first on the outer-shell contribution of Eq. (7.8), p. 145. That contribution is the hydration free energy in liquid water for a distinguished water molecule under the constraint that no inner-shell neighbors are permitted. We will adopt a van der Waals model for that quantity, as in Section 4.1. Thus, we treat first the packing issue implied by the constraint $\prod_j [1 - b_\alpha(j)]$ of Eq. (7.8); then we append a contribution due to dispersion interactions, Eq. (4.6), p. 62. Finally, we include a contribution due to classic electrostatic interactions on the basis of a dielectric continuum model, Section 4.2, p. 67.

For the packing contributions we follow the discussion of Section 4.3. For each configuration sampled from the simulation, 10 000 points were placed randomly in the simulation cell, and the population of water molecules in the defined inner-shell volume calculated. Those sample-averaged frequencies give satisfactory estimates of the p_n that have substantial values, and robust estimates of the moments $\langle n \rangle_0$ and $\langle n(n-1)/2 \rangle_0$. p_0 was obtained by the information theory procedure, and $-kT \ln p_0$ directly gives the packing contribution that is sought. Figure 7.4 shows the $\{p_n\}$ distribution for a cavity of size 3.3 Å.

The distribution $\{x_n\}$ was analyzed in the same way, and is shown in Fig. 7.5 for this particular case. The wings of these distributions are difficult to access,

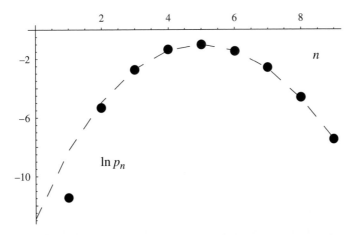

Figure 7.4 $\{p_n\}$ vs. n. The points are the rPBE AIMD simulation results as in Fig. 7.3; the dashed line is the information theory fit to the rPBE results, utilizing the Gibbs default distribution, $\hat{p}_j \propto 1/j!$.

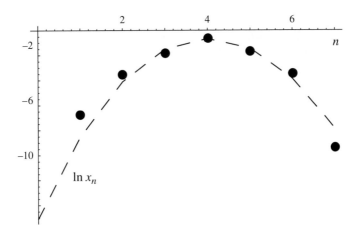

Figure 7.5 $\{x_n\}$ vs. n, as in Fig. 7.4.

and the estimates of p_0 (and x_0) are probably the roughest parts of this example calculation. But since the mean and the second moment seem reliable, the inferred probabilities also appear reasonable.

For the dispersion contribution, we assume that the solute–solvent interaction, in the outer shell, is of the form C/r^6 and that the distribution of water outside the inner shell is uniform. Thus the dispersion contribution is $-4\pi\rho C/(3R^3)$, where for the SPC/E water model, $4\pi\rho C/3$ is $87.3\,\mathrm{kcal\,mol^{-1}\,\mathring{A}^{-3}}$. The electrostatic effects were modeled with a dielectric continuum approach (Yoon and Lenhoff, 1990), using a spherical cavity of radius R. The SPC/E (Berendsen *et al.*, 1987) charge set was used for the water molecule in the center of the cavity.

This procedure can be carried out for observation volumes of different sizes. Of particular interest to us are the sizes 3.0 to 3.4 Å that bracket the minimum in $g_{OO}(r)$ (Fig. 7.3). Figure 7.6 shows the hydration free energy for cavity sizes in this regime. In Fig. 7.6 the minimum for μ^{ex} is obtained for $R = 3.3$ Å. This is consistent with the expectations from $g_{OO}(r)$ (Fig. 7.3).

Using the values for x_0 and p_0 for a cavity of size 3.3 Å, the sum of the chemical (-9.5 kcal mol^{-1}) and packing (8.1 kcal mol^{-1}) contributions is -1.4 kcal mol^{-1}. From scaled-particle theory (Ashbaugh and Pratt, 2006), at 314 K and under saturation conditions, a value of around 6 kcal mol^{-1} is expected for the packing contribution. Our computed value is a bit higher perhaps because the density is a bit higher than that corresponding to saturation conditions at 314 K. Likewise our chemical contribution is expected to be a bit more negative than that expected at 314 K under saturation conditions. But since these effects go in opposite directions, they tend to balance. The results show that the balance between occupancies involved with inner-shell and outer-shell contributions is decisive in establishing an appropriate size of the inner shell. Longer-ranged interactions don't affect that issue significantly, but do contribute importantly to the value of the excess chemical potential found. A more detailed consideration of longer-ranged interactions would be expected to lead to adjustment of the specific net value obtained here.

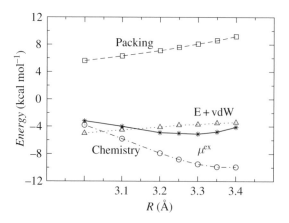

Figure 7.6 Cluster-variation of the hydration free energy of water. The open circles give the chemical contribution, $kT \ln x_0$. The open squares give the packing contribution, $-kT \ln p_0$. The open triangles give the sum of outer-sphere electrostatic and dispersion contributions. The net free energy is shown by a solid line. The minimum is within 1 kcal/mol. of the experimental value of -6.1 kcal/mol.

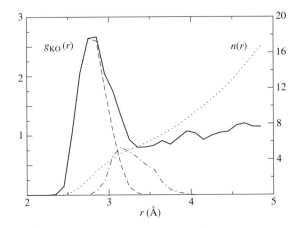

Figure 7.7 The radial distribution function $g_{KO}(r)$ for oxygen atoms about a K^+ ion in liquid water. See Rempe *et al.* (2004). The dashed curve is the contribution to $g_{KO}(r)$ from the nearest four oxygen atoms, and the dashed-dot curve is the contribution from the 5th and 6th nearest oxygen atoms. Notice the lack of definition obtained from the $g_{KO}(r)$ solely because a minimum separating a 1st from a 2nd mean hydration shell is indistinct.

This is a case where an idea of a reasonable inner-shell definition can be obtained from casual observation, Fig. 7.3. But there do exist well-known cases, involving strong interactions, where the observed $g(r)$ does not provide a trivial structural guidance; an example is shown in Fig. 7.7. A quasi-chemical treatment can still be helpful, however.

7.5 Packing in classical liquids

Packing contributions are most evident in the outer-shell contribution to Eq. (7.8) because the indicator function $\Pi_j [1 - b_\alpha (j)]$ constrains that term to the case where no occupancy of the defined inner shell is permitted. Excluded volume effects are the essence of packing contributions. To study those contributions, we consider a solute that does not interact with the solvent at all so that $e^{-\beta \Delta U_\alpha} = 1$. In that case, of course, μ_α^{ex} is zero and we write

$$0 = - \ln \left[1 + \sum_{n \geq 1} \tilde{K}_n \rho^n \right] - \ln \left\langle \! \left\langle \prod_j [1 - b_\alpha (j)] \right\rangle \! \right\rangle_0 \qquad (7.25)$$

where the tilde over the equilibrium coefficients indicates that these coefficients correspond to this specific conceptual case. The rightmost term of Eq. (7.25) gives the contribution to the chemical potential of a solute that perfectly excludes solvent from the region defined by $b_\alpha(j) = 1$ for all j, which we henceforth abbreviate as $\boldsymbol{b} = 1$. Thus, we have the formal result,

$$\beta\mu_\alpha^{\text{ex}}|_{\text{HC}} = \ln\left[1 + \sum_{n \geq 1} \tilde{K}_n \rho^n\right], \qquad (7.26)$$

where 'HC' stands for *hard core*.

This result has a pleasing formal interpretation that develops from Eq. (7.6), p. 145. Successive contributions to the partition function sum of Eq. (7.26) are evidently $\tilde{c}_n/\tilde{c}_0$. The tilde again is a reminder that we are considering the conceptual case on which this argument is founded; $\tilde{c}_n$ is the concentration of those species with n ligands. Noting also that $\tilde{c}_0/\tilde{c}_0 = 1$, the sum is then the total concentration of those species, divided by the concentration of those species with zero (0) ligands, $\tilde{c}_0$. This ratio is the inverse of the probability that a distinguished one of these conceptual cases would have zero (0) ligands. Thus, Eq. (7.26) reduces formally to the well known $\beta\mu_\alpha^{\text{ex}}|_{\text{HC}} = -\ln p(0)$ as in Eq. (4.25), p. 73. In contrast to the more general quasi-chemical formula Eq. (7.8), p. 145, the hard-core case involves no outer-shell term, and includes a change in sign for the inner-shell contribution.

This theory is easy to implement in its most primitive form that uses the approximate value for the equilibrium coefficients obtained by neglecting external medium effects: $\tilde{K}_n \approx \tilde{K}_n^{(0)}$. The quantities $n!\tilde{K}_n^{(0)}$ are then just n-molecule configurational integrals in which all n molecular centers must be *inside* the inner region $\boldsymbol{b} = 1$. Together with the geometric multipliers ρ, Eq. (7.26) then adopts the appearance of a partition function for solvent molecules confined to the region $\boldsymbol{b} = 1$. Self-consistency with the specified solution density $\langle n \rangle_0 = v\rho$, where v is the volume of the $\boldsymbol{b} = 1$ region, can be achieved by augmenting this geometric weighting with a Lagrange multiplier γ that serves as a self-consistent molecular field:

$$\frac{c_n}{c_0} = \tilde{K}_n \rho^n \approx \tilde{K}_n^{(0)} \gamma^n \rho^n. \qquad (7.27)$$

Table 7.1 gives quasi Monte Carlo estimates of the $\tilde{K}_n^{(0)}$ for the case of hard spheres. For the hard-sphere fluid, the predicted distributions x_n for two densities are shown in Figs. 7.8 and 7.9. The predicted occupancy distributions are physically faithful to the data.

Exercise

7.3 Show that

$$n!\tilde{K}_n^{(0)} = \int_v \mathrm{d}^3 r_1 \ldots \int_v \mathrm{d}^3 r_n \left(\prod_{j>i=1}^n e(i, j)\right), \qquad (7.28)$$

Table 7.1 *Quasi-Monte Carlo approximations for $\frac{1}{n}\ln \tilde{K}_n^{(0)}$ (Eq. (7.28)) for unit diameter hard spheres, after Pratt et al. (2001).*

n	1	2	3	4	5	6	7
$\frac{1}{n}\ln \tilde{K}_n^{(0)}$	$\ln\left[\dfrac{4\pi}{3}\right]$	$\dfrac{1}{2}\ln\left[\dfrac{17\pi^2}{36}\right]$	0.189	-0.367	-0.937	-1.53	-2.22

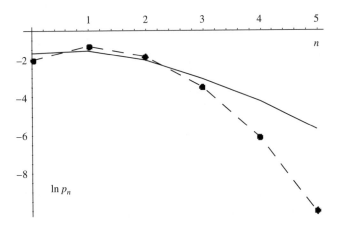

Figure 7.8 For the unit-diameter hard sphere fluid at $\rho = 0.277$, comparison of the Poisson distribution (solid curve) with primitive quasi-chemical distribution Eq. (7.27) (dashed curve). This is the dense gas thermodynamic suggested in Fig. 4.2, p. 74, and the dots are the results of Monte Carlo simulation (Gomez *et al.*, 1999). The primitive quasi-chemical default model depletes the probability of high-n and low-n constellations and enhances the probability near the mode.

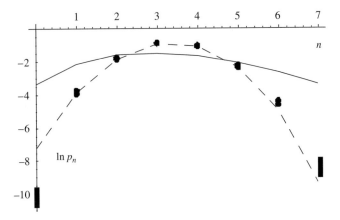

Figure 7.9 As in Fig. 7.8 but for $\rho = 0.8$. The error bars in the wings of the distribution indicate the statistical uncertainty by showing the 67% confidence interval.

where the notation $\int_v \mathrm{d}^3 r_k$ indicates the three-dimensional spatial integral over the volume of a sphere of radius 1, and the integrand factor c(i, j) is zero (0) if $|\mathbf{r}_i - \mathbf{r}_j| < 1$ (overlap) for any (i, j) and one (1) otherwise.

7.6 Self-consistent molecular field for packing

The quantities $\tilde{K}_n(R)$ describe occupancy transformations fully involving the solution neighborhood of the observation volume. These coefficients are known only approximately. Building on the preceding discussion, however, we can go further to develop a self-consistent molecular field theory for packing problems in classical liquids. We discuss here specifically the one component hard-sphere fluid; this discussion follows Pratt and Ashbaugh (2003).

Our primitive quasi-chemical approximation, Eq. (7.27), was

$$\tilde{K}_n \approx \frac{\gamma^n}{n!} \int_v \mathrm{d}\boldsymbol{r}_1 \dots \int_v \mathrm{d}\boldsymbol{r}_n \exp\left[-\sum_{i>j=1}^{n} \beta u(r_{ij})\right]. \tag{7.29}$$

Here $v = 4\pi R^3/3$ is the volume of the observation sphere, $\boldsymbol{b} = 1$. Because of the explicit factors of ρ in Eq. (7.26), γ will approach the thermodynamic excess activity, $\gamma \sim e^{\beta\mu^{\mathrm{ex}}}$, when R is macroscopically large. The integrals of Eq. (7.29) are few-body integrals that can be estimated by Monte Carlo methods as in Table 7.1. A natural extension of the primitive idea is to approximate $\tilde{K}_n(R)$ on the basis of n-molecule configurational integrals that give the low-density limiting quantity, but with inclusion of a molecular field $\beta\phi_{\mathrm{MF}}(\boldsymbol{r})$ as

$$\tilde{K}_n \approx \frac{\gamma^n}{n!} \int_v \mathrm{d}\boldsymbol{r}_1 \dots \int_v \mathrm{d}\boldsymbol{r}_n \exp\left[-\sum_{i=1}^{n} \beta\phi_{\mathrm{MF}}(\boldsymbol{r}_i) - \sum_{i>j=1}^{n} \beta u(r_{ij})\right]$$

$$\equiv \tilde{K}_n^{(0)}[\phi_{\mathrm{MF}}]. \tag{7.30}$$

This molecular field $\beta\phi_{\mathrm{MF}}(\boldsymbol{r})$ describes the effect of the exterior solution on solvent molecules within the observation volume. We will adopt the convention that the molecular field $\beta\phi_{\mathrm{MF}}(\boldsymbol{r})$ be zero at the center of the observation volume. This convention resolves a spatially uniform, additive contribution to $\beta\phi_{\mathrm{MF}}(\boldsymbol{r})$ that would otherwise be ambiguous, and with this convention the Lagrange multiplier γ in this equation may still be recognized as the excess activity as in Eq. (7.29) in the large R limit. The molecular field $\beta\phi_{\mathrm{MF}}(\boldsymbol{r})$, together with the Lagrange multiplier, may be made consistent with the information that the prescribed density of the liquid is uniform within the observation volume. The density profile for the n-molecule case is

$$\rho_n(\boldsymbol{r}) = -\frac{\delta \ln \tilde{K}_n^{(0)}}{\delta \beta\phi_{\mathrm{MF}}(\boldsymbol{r})}[\phi_{\mathrm{MF}}] \tag{7.31}$$

inside the observation volume. Averaging these profiles with respect to the possible occupancies produces the observed density. The consistency sought is then uniformity of the density, see Eq. (6.35), p. 134,

$$-\sum_m p(m)\frac{\delta \ln \tilde{K}_m^{(0)}}{\delta \beta \phi_{\mathrm{MF}}(r)} = \frac{\delta \ln p(0)}{\delta \beta \phi_{\mathrm{MF}}(r)} = -\frac{\delta \beta \mu^{\mathrm{ex}}}{\delta \beta \phi_{\mathrm{MF}}(r)} = \rho, \qquad (7.32)$$

for r inside the observation volume.

Example results are shown in Figs. 7.10–7.14 (Pratt and Ashbaugh, 2003). The self-consistent molecular field was obtained iteratively, including an update of $\tilde{K}_n^{(0)}[\phi_{\mathrm{MF}}]$ by performing additional few-body simulations to evaluate the work associated with turning on the molecular field using thermodynamic integration:

$$\frac{\tilde{K}_n^{(0)}[\phi_{MF}]}{\tilde{K}_n^{(0)}[0]} = \exp\left[-\beta \int_0^1 \left\langle \sum_{j=1}^n \phi_{\mathrm{MF}}(r_j)\right\rangle_\lambda \mathrm{d}\lambda\right],$$

where λ is a coupling parameter, and $\langle\ldots\rangle_\lambda$ indicates averaging over configurations generated under the influence of the molecular field scaled as $\lambda\phi_{\mathrm{MF}}(r)$.

Figure 7.10 shows the self-consistent molecular fields obtained using the procedure described above up to fluid densities of $\rho = 0.9$, just below the hard-sphere

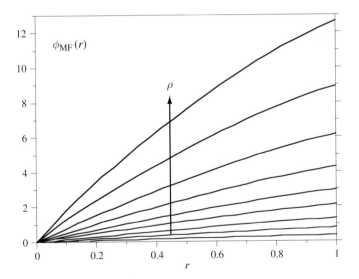

Figure 7.10 The self-consistent molecular field $\beta\phi_{\mathrm{MF}}(r)$ for unit-diameter hard spheres in a spherical observation volume. $r = 0$ is the center of the observation volume, and $r = 1$ is the surface. The curves correspond, from bottom to top, to reduced densities $\rho = 0.1, \ldots, 0.9$, in increments of 0.1 (Pratt and Ashbaugh, 2003).

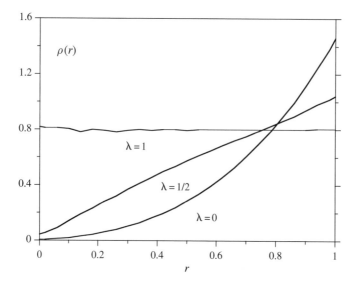

Figure 7.11 Example dependence of the density profile on scaled molecular field $\lambda\phi_{MF}(\boldsymbol{r})$; $\rho = 0.8$. Notice that the density profile is not flat without the full self-consistent mean field (Pratt and Ashbaugh, 2003).

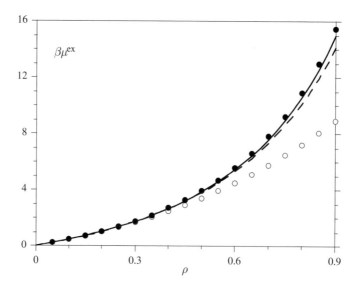

Figure 7.12 Excess chemical potential of the hard-sphere fluid as a function of density. The open and filled circles correspond to the predictions of the primitive quasi-chemical theory and the self-consistent molecular field theory, respectively. The solid and dashed lines are the scaled-particle (Percus–Yevick compressibility) theory and the Carnahan–Starling equation of state, respectively (Pratt and Ashbaugh, 2003).

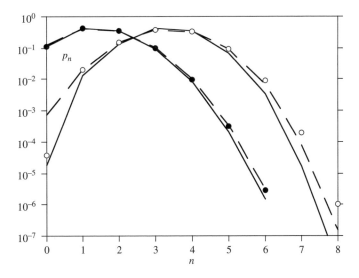

Figure 7.13 Distributions p_n for unit-diameter hard spheres at densities of $\rho = 0.35$ (filled circles) and 0.8 (open circles). The dashed lines are the primitive quasi-chemical theory of Pratt *et al.* (2001), Eq. (7.27), p. 158, and the solid lines correspond to the present MF theory. Note the marked *break-away* of the $n = 0$ point from the primitive quasi-chemical curve (Pratt *et al.*, 2001). The errors on the high n side of these distributions might reflect the fact that the present MF theory doesn't explicitly treat pair correlations. Those correlations enter only through the integrals $\tilde{K}_n^{(0)}[\phi_{MF}]$ (Pratt and Ashbaugh, 2003).

freezing transition. $\phi_{MF}(r)$ is a monotonically increasing function of radial position from the center of the stencil volume to its boundary. This reflects the fact that, in the absence of the molecular field, the hard-sphere particles tend to collect on the surface of the observation volume to minimize their interactions with the other particles (Fig. 7.11). The molecular field makes the boundary repulsive, depletes the surface density, and homogenizes the density within the volume. The magnitude of this repulsive field increases with increasing fluid density.

The predicted hard-sphere chemical potentials as a function of density using the primitive and self-consistent molecular field quasi-chemical theories are compared with the chemical potential from the Carnahan–Starling equation in Fig. 7.12. The primitive theory works well up to $\rho \approx 0.35$, roughly the critical density for Ar and a density region suggested to mark qualitative packing changes in the hard-sphere fluid (Giaquinta and Giunta, 1987). The molecular field theory significantly improves the agreement with the Carnahan–Starling equation over the entire density range. Figure 7.13 shows that the most important deficiencies of the primitive quasi-chemical theory are corrected by the self-consistent molecular field theory. Note that the theory captures the *break-away* of $\ln p(0)$ from the

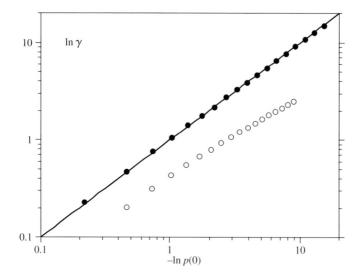

Figure 7.14 Comparison of $\ln \gamma$, with γ the Lagrange multiplier of Eq. (7.30), p. 160, against computed excess chemical potential, $\beta\mu^{ex} = -\ln p(0)$, demonstrating the thermodynamic consistency of these quasi-chemical theories. The open circles are the primitive quasi-chemical theory (Eq. (7.29), p. 160), and the filled circles are the present self-consistent molecular field theory (Pratt and Ashbaugh, 2003).

primitive quasi-chemical prediction at high density. The self-consistent molecular field theory is in close agreement with the scaled-particle (or Percus–Yevick compressibility) theory for the chemical potential.

In addition to achieving a uniform density across the observation volume, the self-consistent molecular field also nearly achieves thermodynamic consistency for the chemical potential. With the choice of an additive constant which makes $\phi_{MF}(r)$ zero in the deepest interior of the observation volume, $\ln \gamma$ should approach $\beta\mu^{ex}$ in the limit of a large R. $\phi_{MF}(r)$ describes an interaction between the interior and the exterior of the observation volume across the intervening boundary. Particularly in the present case of short-ranged interactions, we expect spatial variation of $\phi_{MF}(r)$ to be confined to a surface region. Though an observation volume of radius $R = d$ is evidently not large enough to observe bulk behavior of $\phi_{MF}(r)$ (Fig. 7.10), Fig. 7.14 compares $-\ln p(0)$ and $\ln \gamma$ as determined by the primitive and self-consistent molecular field quasi-chemical theories. While the excess activity evaluated within the primitive theory significantly under-predicts $p(0)$, the self-consistent molecular field theory yields nearly perfect agreement of $\ln \gamma$ and $-\ln p(0)$. At the highest densities, there is a slight disparity between these two quantities, and the calculated values for $\ln \gamma$ are in better agreement with the empirically known $\beta\mu^{ex}$ for the hard-sphere fluid.

The present results address contributions essential to quasi-chemical descriptions of solvation in more realistic cases. An interesting issue is how these packing questions are affected by multiphasic behavior of the solution. In such cases, the self-consistent molecular field ϕ_{MF} should reflect those multiphase possibilities just as it can in pedagogical treatments of nonmolecular models of phase transitions (Ma, 1985).

Exercises

7.4 We have previously shown – see Section 5.6 – that

$$4\pi\rho R^2 G(R) \equiv \frac{\partial\beta\mu^{\mathrm{ex}}}{\partial R}. \tag{7.33}$$

Show that in the present molecular field approximation, the preceding derivative is expressed as

$$4\pi\rho R^2 G(R) \approx \sum_m p(m) \left(\frac{\partial \ln \tilde{K}_m^{(0)}[\phi]}{\partial R} \right). \tag{7.34}$$

7.5 Analyze the derivative required above by considering that the radius R is defined in the first place by a bare field $\beta\phi_0$ that is zero (0) inside the observation volume and ∞ outside. Then the full field encountered with the integrals Eq. (7.30) is $\beta\phi = \beta\phi_0 + \beta\phi_{MF} - \ln\gamma$. Show that the result corresponding to Eq. (7.31) is

$$\frac{\partial \ln \tilde{K}_m^{(0)}[\phi]}{\partial R} = -\int_v \rho_m(r; \beta\phi_{MF}) \frac{\partial\beta\phi(r)}{\partial R} \mathrm{d}^3r.$$

7.6 Show that

$$\frac{\delta\rho_m(r)}{\delta\beta\phi(r')} = -\langle \delta\rho_m(r)\,\delta\rho_m(r')\rangle. \tag{7.35}$$

7.7 Using, then, the averaged quantity

$$-\frac{\delta\rho(r)}{\delta\beta\phi(r')} = \langle \delta\rho(r)\,\delta\rho(r')\rangle \equiv \chi(r, r') \tag{7.36}$$

and

$$-\delta\beta\phi(r) = \int \chi^{-1}(r, r')\,\delta\rho(r')\,\mathrm{d}^3r', \tag{7.37}$$

show that

$$-\frac{\partial\beta\phi(r)}{\partial R} = R^2\rho \int_{|r'|=R_-} \chi^{-1}(r, r')\,\mathrm{d}^2\Omega', \tag{7.38}$$

where the latter integral is over solid angles covering the surface of the ball.

7.8 Finally, introduce $c(\mathbf{r}, \mathbf{r}')$, the Ornstein–Zernike direct correlation function defined by $\chi^{-1}(\mathbf{r}, \mathbf{r}') = \delta(\mathbf{r} - \mathbf{r}')/\rho(\mathbf{r}) - c(\mathbf{r}, \mathbf{r}')$, and derive

$$G(R) = 1 - \int_v c(\mathbf{r}, \mathbf{r}' = \hat{z}R)\, \rho\, \mathrm{d}^3 r \tag{7.39}$$

within the present approximation. In the indicated integral the $\mathbf{r}'$ coordinate is pinned to the sphere surface, and the $\mathbf{r}$ integration is over the interior of the sphere because of Eq. (7.35). The function $c(\mathbf{r}, \mathbf{r}') = c(\mathbf{r}, \mathbf{r}')[\phi]$ is the OZ direct correlation function in the field ϕ including the self-consistent molecular field; thus this is for the case of a uniform density enclosed in a sphere of radius R with no material outside.

7.7 Historical quasi-chemical calculation

Working out the historical quasi-chemical approximation in the present language for the two-dimensional Ising model of a binary solution will give perspective on the developments of this chapter. The model system is depicted in Fig. 7.15. Each site of the lattice possesses a binary occupancy variable, $s_i = \{-1, 1\}$ for the ith site. This will be interpreted so that $s_i = 1$ indicates occupancy of the ith site by one species, e.g. W (water), and $s_i = -1$ then indicates occupancy of that site by the other species, say O (oil). We write

$$n_O \mu_O + n_W \mu_W = \frac{1}{2}(\mu_O + \mu_W)N + \frac{1}{2}(\mu_W - \mu_O)(n_W - n_O). \tag{7.40}$$

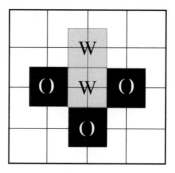

Figure 7.15 Two-dimensional square lattice and figure for the development of a historical quasi-chemical theory. The model assumes that each lattice site will be occupied by either a black ("O") or gray ("W") square. Near-neighbor pairs of the same species contribute $-J < 0$ to the net interaction energy. Near-neighbor pairs of the different species contribute $J > 0$ to the net interaction energy. Our calculation will focus on this five-site figure.

The combination $N = n_W + n_O$ is the nonfluctuating total number of lattice sites and is irrelevant to the statistical issues. We then take

$$\exp\left[\beta J \sum_{nn \text{ pairs}} s_i s_j + \beta(\Delta\mu/2) \sum_i s_i\right], \tag{7.41}$$

with $\Delta\mu = \mu_W - \mu_O$, to be a statistical weight. The generating function

$$\sum \exp\left[\beta J \sum_{nn \text{ pairs}} s_i s_j + \beta(\Delta\mu/2) \sum_i s_i\right], \tag{7.42}$$

then serves as a partition function. Here we use the notation $z = \exp[\beta\Delta\mu/2]$, and $K = \exp[\beta J]$. Considering Fig. 7.15, the partition function for that five-site system is

$$z\left(Kz + \frac{1}{Kz}\right)^4 + \frac{1}{z}\left(\frac{z}{K} + \frac{K}{z}\right)^4. \tag{7.43}$$

Letting x_W be the mole fraction of W (water) occupying the central site, then

$$\frac{x_W}{1 - x_W} = \frac{z\left(Kz + \frac{1}{Kz}\right)^4}{\frac{1}{z}\left(\frac{z}{K} + \frac{K}{z}\right)^4} = z^2\left(\frac{K^2 z^2 + 1}{K^2 + z^2}\right)^4. \tag{7.44}$$

As throughout the book, we notice known concentration factors on the left, leading factors of activities on the right, and our goal is to isolate the dependence of the activities on the concentration.

To make such a toy calculation more accurate in describing a thermodynamically large system, we consider an adjusted activity y as appropriate to replace the factors of z in Eq. (7.44) associated with the peripheral sites, the *ligands* of the complex. This is intended to compensate for the fact that the cluster is amputated there. Then

$$\frac{x_W}{1 - x_W} = z^2\left(\frac{K^2 y^2 + 1}{K^2 + y^2}\right)^4. \tag{7.45}$$

To evaluate the mean-field activity y, apply the physically equivalent treatment to one of the peripheral sites:

$$\frac{x_W}{1 - x_W} = \frac{y\left[Kz\left(Ky + \frac{1}{Ky}\right)^3 + \frac{1}{Kz}\left(\frac{y}{K} + \frac{K}{y}\right)^3\right]}{\frac{1}{y}\left[\frac{z}{K}\left(Ky + \frac{1}{Ky}\right)^3 + \frac{K}{z}\left(\frac{y}{K} + \frac{K}{y}\right)^3\right]}$$

$$= y^2\left[\frac{\left(K^2 + y^2\right)^3 + K^2\left(1 + K^2 y^2\right)^3 z^2}{K^2\left(K^2 + y^2\right)^3 + \left(1 + K^2 y^2\right)^3 z^2}\right]. \tag{7.46}$$

The activity y can be algebraically eliminated between Eqs. (7.45) and (7.46) yielding

$$
\sqrt{z} = \left(\frac{x_W}{1 - x_W}\right)^{1/4}
$$
$$
\times \left(\frac{K^2 (1 - x_W)^{3/4} + x_W^{3/4} \left(K^4 (1 - x_W) + x_W\right) \sqrt{z}}{(1 - x_W)^{3/4} \left(1 + (-1 + K^4) x_W\right) + K^2 x_W^{3/4} \sqrt{z}}\right). \quad (7.47)
$$

Finally, solving these equations for z gives

$$
\beta \Delta \mu = \ln\left(\frac{x_W}{1 - x_W}\right)
$$
$$
+ 4 \ln\left(\frac{-1 + 2 x_W + \sqrt{(1 - 2 x_W)^2 + 4 K^4 (1 - x_W) x_W}}{2 K^2 x_W}\right). \quad (7.48)
$$

This predicts a critical point for phase separation at $K^2 = 2$ and $x_w = 1/2$. Exact results for the two-dimensional Ising lattice gas model are $K^2 = \frac{1}{\sqrt{2}-1} \approx 2.4$ and $x_w = 1/2$ (Hill, 1987, Section 44).

7.8 Explicit–implicit solvent models

In view of the results of the previous two sections, we can carry the practical development of quasi-chemical approximations further. We will initially consider cases for which the interactions are fundamentally short-ranged. For the principal contrary example, ions, the outer-shell term would represent the *Born* contribution because it describes a hard ion stripped of any inner-shell ligands. But the balance of treatments of long-ranged interactions requires specific subsequent consideration.

With that motivation and restriction, we identify the outer-shell term as an initial packing contribution

$$
\left\langle\!\!\left\langle e^{-\beta \Delta U_\alpha} \prod_j [1 - b_\alpha (j)]] \,|\, \mathcal{R}^n \right\rangle\!\!\right\rangle_0 = \left\langle\!\!\left\langle \prod_j [1 - b_\alpha (j)] \,|\, \mathcal{R}^n \right\rangle\!\!\right\rangle_0. \quad (7.49)
$$

This is a packing contribution of the type analyzed previously. Adopting the simplest of the preceding results, Eq. (7.30) p. 160, directly we have

$$
\left\langle\!\!\left\langle e^{-\beta \Delta U_\alpha} \prod_j [1 - b_\alpha (j)] \,|\, \mathcal{R}^n \right\rangle\!\!\right\rangle_0 \approx \frac{1}{1 + \sum_{n \geq 1} \tilde{K}_n^{(0)} [\phi_{MF}] \gamma^n \rho^n}. \quad (7.50)
$$

Returning to consider the *inner-shell* contribution, following a preceding section, the natural idea is to exploit the same self-consistent molecular field to approximate

$$K_n \approx \gamma^n K_n^{(0)} [\phi_{\mathrm{MF}}] \qquad (7.51)$$

so that

$$\mu_\alpha^{\mathrm{ex}} (\mathcal{R}^n) \approx -kT \ln \left\{ \frac{1 + \sum_{n \geq 1} K_n^{(0)} [\phi_{\mathrm{MF}}] \gamma^n \rho^n}{1 + \sum_{n \geq 1} \tilde{K}_n^{(0)} [\phi_{\mathrm{MF}}] \gamma^n \rho^n} \right\}. \qquad (7.52)$$

Substituting

$$K_n^{(0)} [\phi_{\mathrm{MF}}] = \left(\frac{K_n^{(0)} [\phi_{\mathrm{MF}}]}{\tilde{K}_n^{(0)} [\phi_{\mathrm{MF}}]} \right) \tilde{K}_n^{(0)} [\phi_{\mathrm{MF}}] \qquad (7.53)$$

in the numerator gives

$$\frac{1 + \sum_{n \geq 1} K_n^{(0)} [\phi_{\mathrm{MF}}] \gamma^n \rho^n}{1 + \sum_{n \geq 1} \tilde{K}_n^{(0)} [\phi_{\mathrm{MF}}] \gamma^n \rho^n} = \left\langle \left(\frac{K_n^{(0)} [\phi_{\mathrm{MF}}]}{\tilde{K}_n^{(0)} [\phi_{\mathrm{MF}}]} \right) | \mathcal{R}^n \right\rangle_{\mathrm{GC}}, \qquad (7.54)$$

adapting the notation of Eq. (3.17), p. 40; GC indicates a *grand canonical* average for a system confined to the volume $b = 1$, here utilizing the activity $\gamma \rho$. The quantities being averaged depend on the molecular field ϕ_{MF}. The numerator and denominator of the ratio appearing on the right side of Eq. (7.54) are configurational integrals involving the same coordinates. The integrands differ only in the Boltzmann factor of solute–solution interactions. Thus, we can rewrite Eq. (7.52) as

$$\mu_\alpha^{\mathrm{ex}} (\mathcal{R}^n) \approx -kT \ln \left\langle e^{-\beta \Delta U_\alpha} | \mathcal{R}^n \right\rangle_{\mathrm{GC}} [\phi_{MF}]. \qquad (7.55)$$

This notation $[\phi_{MF}]$ emphasizes that this is a functional of ϕ_{MF}, the enclosed material being subject to the molecular field ϕ_{MF}.

Evaluation of the partition function Eq. (7.55) typically would not be a trivial task, but Chapter 5, p. 100, can be brought to bear because this is a standard form.

Correspondence to multigaussian models

The result Eq. (7.54) is also remarkable because it takes the form anticipated intuitively by the multigaussian theory Eq. (4.23), p. 71. The observation that

$$\tilde{p} (s) = \frac{\tilde{K}_s^{(0)} [\phi_{\mathrm{MF}}] \gamma^s \rho^s}{1 + \sum_{n \geq 1} \tilde{K}_n^{(0)} [\phi_{\mathrm{MF}}] \gamma^n \rho^n} \qquad (7.56)$$

makes that clear. Here the $\tilde{p}(s)$ are the occupancy probabilities for the $b = 1$ region. Even for packing problems, this $b = 1$ region might be larger than the excluded volume; it might naturally include an anticipated first-shell region. We can then generalize Eq. (7.54) beyond the thermodynamic potential distribution theorem to

$$\mathcal{P}_{\alpha}^{(0)}(\varepsilon) = \sum_{s \geq 0} \mathcal{P}_{\alpha}^{(0)}(\varepsilon|s)\tilde{p}(s), \tag{7.57}$$

which can be compared to Eq. (4.23), p. 71. For excluded-volume problems, this expression would be written

$$p(n) = \sum_{s \geq 0} p(n|s)\tilde{p}(s). \tag{7.58}$$

This relation says that the probability of n solvent centers occupying the excluded volume is the product of two probabilities: the probability that the larger ($b = 1$) region has s solvent centers multiplied by the conditional probability that the observation volume has n centers when the larger region has s centers. For the packing problems the approximation obtained above amounts to $p(n = 0|s) \approx K_s^{(0)}[\phi_{\text{MF}}]/\tilde{K}_s^{(0)}[\phi_{\text{MF}}]$ with a quasi-chemical approximation Eq. (7.56) for $\tilde{p}(s)$. If a gaussian model for $p(n|s)$ were satisfactory for Eq. (7.58), then this would be a multigaussian model.

Implicit solvent models

The solvent is a necessary part of the physical problem for computational studies of larger-scale structures in solution. But often the solvent is of secondary interest. Therefore, there has been extended attention to *implicit solvent* models for those computational studies, models that provide the proper statistical description of the macromolecule but without the solvent explicitly present (Roux and Simonson, 1999). Equation (3.38), p. 45, provides a fundamental basis for implicit solvent models.

There seems to be a well-developed folklore that judicious explicit inclusion of a small number of solvent (water) molecules can dramatically improve the accuracy of implicit hydration models (Gilson *et al.*, 1997). An important physical observation is that an appropriate inclusion of an inner shell only can capture most of the effects of the solvent on the solute of interest (Beglov and Roux, 1994; 1995; Bizzarri and Cannistraro, 2002). The quasi-chemical approach is the theory for inclusions of that sort.

For macromolecules in solution, treatment of conformational fluctuations and shape variations, which might be treated by implicit solvent models, is the most important issue for simulations with periodic boundary conditions. The present quasi-chemical approach permits shape fluctuations if the inner shell is defined with respect to atomic sites of the macromolecular solutes that become more or less accessible as the conformation of the solute changes in sampling $s_\alpha^{(0)}(\mathcal{R}^n)$. The volume of the inner-shell region can then change, and consequently the occupancies also. In addition to suggesting *explicit–implicit* models, the quasi-chemical theory permits well-defined investigation of further corrections to simple models.

8

Developed Examples

This chapter takes up a sequence of examples of the concepts discussed above. Each example is in the nature of a seminar. Initial examples of quasi-chemical calculations are presented in Sections 7.3, 7.4, and 7.6.

8.1 Polymers

Polymer solutions figure in a vast array of practical materials and processes in the modern world. Ideas about polymers are also relevant to understanding solutions of DNA, proteins, polysaccharides, and other solutions of biological interest. Because of the size and complexity of the chain molecule solutes, polymer solutions present challenging problems in solution theory, and a great deal of work has been directed toward a theoretical understanding of these solutions over the last century.

Chemical potentials and the equation of state

Okamoto (1993) and Escobedo and dePablo (1995) relate the pressure to the excess chemical potentials on the basis of a Gibbs–Duhem integration. We integrate

$$\mathrm{d}p = \sum_\alpha \rho_\alpha \mathrm{d}\mu_\alpha \tag{8.1}$$

with respect to volume $\mathcal{V}$ with temperature T and particle numbers n_α fixed. The ideal gas functions obey this relation so we can subtract that relation for the ideal functions, and use the same expression

$$\mathrm{d}p^{\mathrm{ex}} = \sum_\alpha \rho_\alpha \mathrm{d}\mu_\alpha^{\mathrm{ex}} \tag{8.2}$$

for the excess quantities. It is convenient to integrate by parts

$$
d\left(p^{\text{ex}} - \sum_\alpha \rho_\alpha \mu_\alpha^{\text{ex}}\right) = -\sum_\alpha \mu_\alpha^{\text{ex}} n_\alpha \, d\left(\frac{1}{\mathcal{V}}\right). \tag{8.3}
$$

Note that excess properties are zero in the limit of infinite volume $\mathcal{V}$, so that

$$
\frac{\beta p \mathcal{V}}{N} = 1 + \sum_\alpha x_\alpha \left(\beta \mu_\alpha^{\text{ex}} - \int_0^1 \beta \mu_\alpha^{\text{ex}}(\lambda \rho) \, d\lambda\right), \tag{8.4}
$$

where $\mu_\alpha^{\text{ex}}(\lambda \rho)$ are the excess chemical potentials evaluated at the densities $\rho = \{\rho_1 \ldots\}$ where the density is scaled by λ. Equation (8.4), the *osmotic equation of state*, provides a direct connection between the potential distribution theorem and the equation of state for any molecular mixture.

Exercises

8.1 The osmotic pressure is the pressure difference between two systems that equilibrate with respect to transfer of one component, the solvent, but not other components, perhaps a polymeric solute. Following Eq. (8.4), derive a general formula for the osmotic pressure.

8.2 Derive an explicit formula for the osmotic pressure, correct for the lowest solute concentrations.

8.3 On the basis of the general formulae obtained above, see how far you can get in evaluating the osmotic pressure at the next higher order of solute concentration, discussing how your results might be evaluated on the basis of molecular structural information.

Flory–Huggins theory

Our discussion here explores active connections between the potential distribution theorem (PDT) and the theory of polymer solutions. In Chapter 4 we have already derived the Flory–Huggins model in broad form, and discussed its basis in a van der Waals model of solution thermodynamics. That derivation highlighted the origins of composition, temperature, and pressure effects on the Flory–Huggins interaction parameter. We recall that this theory is based upon a van der Waals treatment of solutions with the additional assumptions of zero volume of mixing and more technical approximations such as Eq. (4.45), p. 81. Considering a system of a polymer (p) of polymerization index M dissolved in a solvent (s), the Flory–Huggins model is

$$
\frac{\beta \Delta \mathcal{G}_{\text{mix}}}{\bar{\rho}_s \mathcal{V}} = \phi_s \ln \phi_s + \frac{(1 - \phi_s)}{M} \ln(1 - \phi_s) + \chi_{\text{sp}} \phi_s (1 - \phi_s). \tag{8.5}
$$

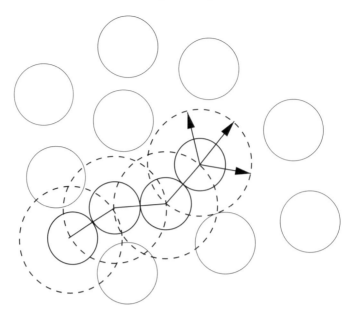

Figure 8.1 The process of computing the incremental chemical potential involves adding one extra segment to an $M-1$ segment chain moving in the solvent. The tangent hard sphere model of a $(M-1)$-mer $(M=5)$ is shown here. The dashed circles enclose the volume excluded to the centers of the solvent spheres.

See Eq. (4.56), p. 85. The chemical potentials are correspondingly

$$\beta\mu_s = \beta\bar{\mu}_s + \ln\phi_s + \left(1 - \frac{1}{M}\right)(1 - \phi_s) + \chi_{sp}(1 - \phi_s)^2 \qquad (8.6a)$$

$$\beta\mu_p = \beta\bar{\mu}_p + \ln(1 - \phi_s) - (M-1)\phi_s + M\chi_{sp}\phi_s^2. \qquad (8.6b)$$

(See Eq. (4.62), p. 87). If χ_{sp} is large and positive, a loop appears as in Fig. 4.7, p. 86, indicating a separation into a polymer-rich phase and a solvent-rich phase.

For a polymer blend of two components,

$$\frac{\beta\Delta\mathcal{G}_{mix}}{(M_1\bar{\rho}_1 + M_2\bar{\rho}_2)\mathcal{V}} = \left[\frac{\phi_1}{M_1}\ln\phi_1 + \frac{\phi_2}{M_2}\ln\phi_2 + \chi_{12}\phi_1\phi_2\right] \qquad (8.7)$$

is the corresponding free energy of mixing. Here M_1 and M_2 are the polymerization indices on the scale of the size attributed to a monomer. It is clear that if M_1 and M_2 are large, the interaction term dominates the free energy of mixing. Since for many polymer systems $\chi_{12} > 0$, it is often the case that mixtures of two different polymers phase separate due to the interaction.

Consider computing the excess chemical potential for a chain molecule composed of M monomers, denoting such a molecule type by α_M. The longer the chain, the more difficult the computation. It is easier to compute the *incremental* chemical potential for addition of a single segment. Following the notation discussed in Section 1.3, p. 16,

$$e^{-\beta \mu_{\alpha_M}^{ex} (\mathcal{R}^n)} = \left\langle e^{-\beta \Delta U_{\alpha_M}} | \mathcal{R}^n \right\rangle_0 . \tag{8.8}$$

To assist in the statistical evaluation on the right side of this relation, we choose $e^{-\beta \Delta U_{\alpha_{M-1}}}$ as an importance function so that

$$e^{-\beta \mu_{\alpha_M}^{ex} (\mathcal{R}^n)} = e^{-\beta \mu_{\alpha_{M-1}}^{ex} (\bar{\mathcal{R}}^n)} \left\langle e^{-\beta \left(\Delta U_{\alpha_M} - \Delta U_{\alpha_{M-1}} \right)} | \mathcal{R}^n \right\rangle_{M-1} . \tag{8.9}$$

Here $\bar{\mathcal{R}}^n$ is the subset of all $\mathcal{R}^n$ that specifies the configuration of the α_{M-1} molecule. The notation $\langle \ldots | \mathcal{R}^n \rangle_{M-1}$ indicates the expected value conditional on the configuration $\mathcal{R}^n$ and with the importance function $e^{-\beta \Delta U_{\alpha_{M-1}}}$ associated with the $(M-1)$-mer. We compute, then, the average Boltzmann factor for the incremental energy required to add the Mth segment. It is clear from the experience suggested by Fig. 1.10, p. 20, and the surrounding discussion, that this type of restriction can achieve a surprising simplification. The picture of adding a segment to the end of a chain to obtain the incremental excess chemical potential is close in spirit to the Kirkwood–Salsburg statistical theory presented in Section 6.1, p. 123.

Exercise

8.4 Why don't the Flory–Huggins approximations for the chemical potentials Eqs. (8.6) require any evaluation of the internal partition function of a single chain molecular, q_p^{int}?

Monte Carlo methods for chain molecules

Molecule-scale statistical simulations of chain molecular solutions are more specialized than simulations of small-molecule solutions. Here we discuss some of the special issues that come up. Dynamical simulations have also been pursued, and those results typically have been satisfactory, though fundamental questions have been raised (Madras and Sokal, 1987; Sokal, 1995).

Sampling configurations of an isolated chain molecule

Because the thermodynamic parameter is expressed as

$$e^{-\beta \mu_{\alpha_M}^{ex}} = \mathcal{V}^{-1} \int \left\langle e^{-\beta \Delta U_{\alpha_M}} | \mathcal{R}^n \right\rangle_0 s_{\alpha_M}^{(0)} (\mathcal{R}^n) \mathrm{d} (\mathcal{R}^n) , \tag{8.10}$$

in terms of the statistical problem of Eq. (8.8), let's first consider sampling $s_{\alpha_M}^{(0)}(\mathcal{R}^n)$. An idea for a practical calculation is to sample $s_{\alpha_M}^{(0)}(\mathcal{R}^n)$, then estimate the required integrand, Eq. (8.8), either directly on that basis, or incrementally as with Eq. (8.9).

One charming idea for sampling $s_{\alpha_M}^{(0)}(\mathcal{R}^n)$ is *inversely restricted sampling* (IRS) (Hammersley and Morton, 1954), or *Rosenbluth sampling* (Rosenbluth and Rosenbluth, 1955); see also Hammersley and Handscomb (1964, Section 10.3). The idea, physically expressed, is to design a growth algorithm for serial construction of a chain configuration, see Fig. 8.2. This algorithm should be simple enough to obtain nonstatistically the normalized probability distributions for configurations produced on that basis; let's call that distribution $1/\Omega(\mathcal{R}^n)$ for chain configuration $\mathcal{R}^n$. Such a growth mechanism needn't produce configurations in proportions consistent with statistical equilibrium at temperature T, and they don't do that in the typical case (Widom, 1966; McCrackin, 1972). But integrals such as

$$\int C(\mathcal{R}^n)\,\mathrm{d}\mathcal{R}^n = \int C(\mathcal{R}^n)\,\Omega(\mathcal{R}^n)\left(\frac{1}{\Omega(\mathcal{R}^n)}\right)\mathrm{d}\mathcal{R}^n \qquad (8.11)$$

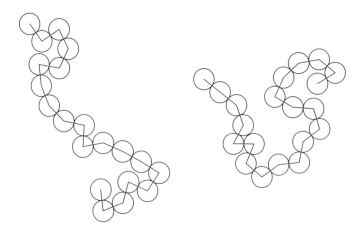

Figure 8.2 Examples of freely-jointed, tangent hard disk chains ($M = 20$) produced by an inversely restricted sampling method. In this case, the angle ϕ_j of the jth bond was chosen randomly in the fraction of 2π available without overlap of the $(j+1)$th disk with any previously placed disk, $1\ldots j$. The ratio of weights Ω (left)$/\Omega$ (right) ≈ 0.76, so the left configuration is more probable than the right configuration, whereas the Boltzmann distribution assigns equal probabilities to these configurations; see McCrackin (1972). Thus, in this case if the left configuration were the proposed, or trial configuration to succeed the right configuration – $r > 1$ in Eq. (8.14) – it would be accepted with probability one. If the right configuration were the trial configuration and the left configuration the current state, then the right configuration would be accepted as the next state with probability 0.76, and the left one with probability 0.24.

can be estimated in the usual way

$$\int C\left(\mathcal{R}^n\right)\mathrm{d}\mathcal{R}^n \propto E\left(\frac{1}{m}\sum_{k=1}^{m}\Omega\left([\mathcal{R}^n]_k\right)C\left([\mathcal{R}^n]_k\right)\right), \qquad (8.12)$$

provided the conventional multiply-and-divide is permitted; see Section 5.1, p. 100. The proportionality of Eq. (8.12) reflects an acknowledged lack of detail with respect to dimensional issues here. The sample for Eq. (8.12) is generated with the growth algorithm. With this background, averages corresponding to statistical equilibrium at temperature T would be estimated from

$$\mathcal{V}^{-1}\int C\left(\mathcal{R}^n\right)s_{\alpha_M}^{(0)}\left(\mathcal{R}^n\right)\mathrm{d}\left(\mathcal{R}^n\right)$$

$$=\frac{E\left(\sum_{k=1}^{m}\Omega\left([\mathcal{R}^n]_k\right)\mathrm{e}^{-\beta U_{\alpha_M}^{(1)}\left([\mathcal{R}^n]_k\right)}C\left([\mathcal{R}^n]_k\right)\right)}{E\left(\sum_{k=1}^{m}\Omega\left([\mathcal{R}^n]_k\right)\mathrm{e}^{-\beta U_{\alpha_M}^{(1)}\left([\mathcal{R}^n]_k\right)}\right)}. \qquad (8.13)$$

Because of the ratio, the estimate Eq. (8.13) is biased, in the conventional sense of Section 5.5, p. 118, though the estimate Eq. (8.12) is unbiased in that conventional sense. Such an approach can work in cases of interest, but the characteristic difficulty is that the weights in the summands of Eq. (8.13) grow exponentially with M; and concurrently "... the weights are almost certain to get out of hand, a few of them being much larger than all the rest." (Hammersley and Handscomb, 1964, Section 10.3).

For the purposes of sampling the normalized distribution $s_{\alpha_M}^{(0)}\left(\mathcal{R}^n\right)$, in contrast with estimation of integrals such as Eq. (8.12), the procedure above can be improved by utilizing the Metropolis rejection idea (Kalos and Whitlock, 1986, Section 3.7). Suppose that a configuration $[\mathcal{R}^n]_k$ is in hand. Its probability density on the basis of the growth algorithm is $1/\Omega\left([\mathcal{R}^n]_k\right)$. A typical choice for this probability density is the Boltzmann factor for an added segment in a test direction divided by the sum of the Boltzmann factors over a randomly chosen set of test directions; the product of all of these factors for the grown chain yields the final $1/\Omega\left([\mathcal{R}^n]_k\right)$. Now on the basis of the growth algorithm generate another configuration, $[\mathcal{R}^n]_{k'}$ with probability density $1/\Omega\left([\mathcal{R}^n]_{k'}\right)$, and view this configuration as a proposed, or *trial* configuration in the sense of the Metropolis Monte Carlo algorithm. Thus, if the ratio of probabilities

$$r=\left[\frac{\mathrm{e}^{-\beta U_{\alpha_M}^{(1)}\left([\mathcal{R}^n]_{k'}\right)}\Big/\mathrm{e}^{-\beta U_{\alpha_M}^{(1)}\left([\mathcal{R}^n]_k\right)}}{\left(\frac{1}{\Omega\left([\mathcal{R}^n]_{k'}\right)}\right)\Big/\left(\frac{1}{\Omega\left([\mathcal{R}^n]_k\right)}\right)}\right] \qquad (8.14)$$

is greater than one, $r>1$, then choose the $(k+1)$th state to be $[\mathcal{R}^n]_{k'}$. If $r\leq 1$, then choose the $(k+1)$th state to be $[\mathcal{R}^n]_{k'}$ with probability r, or the $(k+1)$th state to be $[\mathcal{R}^n]_k$ with probability $1 - r$. This choice for the Metropolis selection criterion assures detailed balance, and thus the generation of states that sample the Boltzmann distribution.

Notice that this latter procedure produces two data streams, one the IRS stream and the other the Boltzmann data stream. Both data streams are useful, and the ratios of the distributions obtained from these data streams are useful also.

The above discussion has focused on generating free chain configurations to sample the distribution $s_{\alpha_M}^{(0)}(\mathcal{R}^n)$. These configurations can then be utilized to compute the excess chemical potential for short chains directly, Eq. (8.8), or the incremental excess chemical potential (Kumar *et al.*, 1996) for longer chains on the basis of Eq. (8.9). A direct extension of the IRS sampling scheme to include interactions with solvent and other polymers in the solution is the *configurational bias Monte Carlo* method; for reviews and applications to chain-molecule phase equilibria, see Mooij and Frenkel (1994), dePablo and Escobedo (1999) and Siepmann (1999).

Exercises

8.5 Reconsider the Metropolis rejection Monte Carlo method (Kalos and Whitlock, 1986, Section 3.7), and derive the acceptance probability Eq. (8.14) on the basis of the traditional detailed balance specification.

8.6 Use the potential distribution theorem and the above discussion to derive an expression for the excess chemical potential computed using the IRS idea.

A generalized Flory–Huggins theory

To address the limitations of ancestral polymer solution theories, recent work has studied specific molecular models – the tangent hard-sphere chain model of a polymer molecule – in high detail, and has developed a generalized Flory theory (Dickman and Hall (1986); Yethiraj and Hall, 1991). The justification for this simplification is the van der Waals model of solution thermodynamics, see Section 4.1, p. 61: attractive interactions that stabilize the liquid at low pressure are considered to have weak structural effects, and are included finally at the level of first-order perturbation theory. The packing problems remaining are attacked on the basis of a hard-core model reference system.

Thus, we first consider Eq. (8.10) for hard-core chain models, specifically tangent hard-sphere chain models (Dickman and Hall (1986); Yethiraj and Hall, 1991). Models and theories of the packing problems associated with hard-core molecules have been treated in Sections 4.3, 6.1, 7.5, and 7.6. We recall

$$e^{-\beta\tilde{\mu}_{\alpha_M}^{ex}(\mathcal{R}^n)} = p_{\alpha_M}(0|\mathcal{R}^n). \tag{8.15}$$

(See Eq. (4.25), p. 73.) $p_{\alpha_M}(0|\mathcal{R}^n)$ is the probability that a stencil outlining the excluded volume of the α_M polymer, with conformation sampled from the isolated molecular potential energy surface, would have zero (0) occupancy.

The simplest view for our present problem is based upon the Flory–Huggins scaling $\tilde{b}_{sp}^{(2)}/\tilde{b}_{ss}^{(2)} \approx M$, p. 81. Thus, we expect that $\beta\mu_{\alpha_M}^{ex} \sim M$. Consideration of the incremental free energy changes upon addition of a terminal site suggests how this might work:

$$p_{\alpha_M}(0|\mathcal{R}^n) = p_\alpha(0|\mathcal{R}^n) \prod_{M'=2}^{M'=M} \left(\frac{p_{\alpha_{M'}}(0|\mathcal{R}^n)}{p_{\alpha_{M'-1}}(0|\mathcal{R}^n)} \right). \tag{8.16}$$

Each of the factors on the right side of Eq. (8.16) reflects a statistical assessment of the possibilities for insertion of a single monomer.

The conventional idea for making this formal expression tractable is analogous to the Ursell development, p. 126.

$$\ln p_{\alpha_M}(0|\mathcal{R}^n) = \sum_{M'=1}^{M'=M} \omega_{M'}^{(1)}(\mathcal{R}^n) + \sum_{pairs} \omega_{M'M''}^{(2)}(\mathcal{R}^n) + \cdots \tag{8.17}$$

It is simplest, though not necessary, to restrict the pair sum to just chemically adjacent pairs, so we do that for the present discussion. The prescription for determining the functions $\omega^{(j)}(\mathcal{R}^n)$ is that they should make the development correct if $M = j$. Thus

$$\omega_{M'}^{(1)}(\mathcal{R}^n) = \ln p_\alpha(0|\mathcal{R}^n), \tag{8.18a}$$

$$\omega_{M'M'+1}^{(2)}(\mathcal{R}^n) = \ln p_{\alpha_2}(0|\mathcal{R}^n) - 2\ln p_\alpha(0|\mathcal{R}^n). \tag{8.18b}$$

This generates the sequence of approximations

$$p_{\alpha_M}(0|\mathcal{R}^n) \approx p_\alpha(0|\mathcal{R}^n)^M, \tag{8.19a}$$

$$p_{\alpha_M}(0|\mathcal{R}^n) \approx p_\alpha(0|\mathcal{R}^n) \left(\frac{p_{\alpha_2}(0|\mathcal{R}^n)}{p_\alpha(0|\mathcal{R}^n)} \right)^{M-1}, \tag{8.19b}$$

the latter approximation being similar to the *son-of-superposition* approximation (Chae *et al.*, 1969). More subtly, Dickman and Hall (1986) propose the sequence of approximations

$$p_{\alpha_M}(0|\mathcal{R}^n) \approx p_\alpha(0|\mathcal{R}^n)^{v_e(M)/v_e(1)}, \tag{8.20a}$$

$$p_{\alpha_M}(0|\mathcal{R}^n) \approx p_\alpha(0|\mathcal{R}^n) \left(\frac{p_{\alpha_2}(0|\mathcal{R}^n)}{p_\alpha(0|\mathcal{R}^n)} \right)^{\frac{v_e(M)-v_e(1)}{v_e(2)-v_e(1)}}, \tag{8.20b}$$

revising the simple exponents to reflect the fact that the exposed region of influence of successive spheres overlap, as in Fig. 1.10, p. 20; the revised exponents eventually increase linearly with M. The equation of state for the hard dumb-bell fluid (Tildesley and Streett, 1980) can be used to implement the pair theory. The theory has been extended to incorporate a square-well attractive potential (Yethiraj and Hall, 1991).

These approximations can then be used in the osmotic equation of state to obtain the compressibility factor. Monte Carlo simulations using the above-discussed Monte Carlo techniques have been performed to assess the approximations inherent in the generalized Flory theory of hard-core chain systems. This theory does quite well in predicting the equations of state of hard-core chains at fluid densities. The question then arises, why does it do so well since the theory typically only incorporates information from a dimer fluid as a reference state?

The clearest answer to this question comes from comparisons of the simulations with the generalized Flory theory. This is because of the direct connection to the PDT through the cavity probabilities. The simulations have shown (Escobedo and dePablo, 1995; Kumar *et al.*, 1996) that the weakest approximation is that of assuming the cavity probability of a monomer in the M-mer fluid is the same as that for a monomer in a monomer fluid. The generalized Flory theory underestimates this cavity probability (thus overestimating the excess chemical potential) by up to 40%. The error is in turn balanced by errors in the conditional cavity probabilities for addition of succeeding segments, leading to overall good agreement. Further development of the QCA approach to chain molecules will be interesting since this theory incorporates a more detailed picture of the local structure around the chain segments. This section has only touched on the relevance of the potential distribution theorem to polymer science.

Exercise

8.7 Write out the general term following from the Ursell expansion, and describe what to do if you have results for an *n*-alkane in water.

8.2 Primitive hydrophobic effects

"No one has yet proposed a quantitative theory of aqueous solutions of nonelectrolytes, and such solutions will probably be the last to be understood fully." (Rowlinson and Swinton, 1982).

Hydrophobic and hydrophilic are categories of hydration effects in aqueous liquids. Classical ions such as Na^+ or polar molecules such as urea $[(NH_2)_2 CO]$ are simply recognized hydrophilic solutes. In contrast, the interactions of hydrophobic solutes or groups with water molecules do not display classic electrostatic or specific chemical interactions. Primitive hydrophobic solutes are inert gases and simple hydrocarbons that are sparingly soluble in water. Nevertheless, much of the interest in hydrophobic effects is associated with more complex solutes that contain both hydrophobic and hydrophilic moieties. Surfactant molecules, for example the decanoate anion, include both hydrophobic and hydrophilic parts and are called amphiphilic.

Solutions containing amphiphiles often show exotic behaviors. The assembly of micelles and bilayer membranes is associated with the bifunctional character of the species that compose them. These structures attempt to sequester hydrophobic groups away from the aqueous environment while still satisfying tendencies of hydrophilic groups for contact with water. Many amino acids and peptides are amphiphilic molecules. Protein molecular structure, function, and aggregation has provided an important motivation for the study of hydrophobic effects. It is widely believed that hydrophobic interactions drive protein folding by providing a nonspecific, cohesive stabilization of structures that successfully satisfy the contrasting hydration requirements of hydrophilic and hydrophobic molecular parts.

Hydrophobic effects are thus of practical interest. If we accept the goal of a simple, physical, molecularly valid explanation, then hydrophobic effects have also proved conceptually subtle. The reason is that hydrophobic phenomena are not tied directly to a simple dominating interaction as is the case for hydrophilic hydration of Na^+, as an example. Instead hydrophobic effects are built up more collectively. In concert with this indirectness, hydrophobic effects are viewed as entropic interactions and exhibit counterintuitive temperature dependencies. An example is the cold denaturation of globular proteins. Though it is believed that hydrophobic effects stabilize compact protein structures and proteins denature when heated sufficiently, it now appears common for protein structures to unfold upon appropriate cooling. This entropic character of hydrophobic effects makes them more fascinating and more difficult.

The strictly hydrophobic case is one in which ΔU involves no classic electrostatic interactions, no hydrogen bonding, and no other chemical or associative interactions; ΔU is of the van der Waals type. In the extreme model, ΔU involves only hard-core repulsions preventing overlap of the van der Waals volume of any solution constituents with the van der Waals volume of a solute molecule. This approach is consistent with the view that dissolving a solute can be considered as a two-step process. First, a cavity for the solute is created and then the solute is placed in this cavity. Final contributions from other interactions are typically interesting, but are not addressed at this stage.

Hard-core models of solute–water interactions serve as a valuable reference point for two reasons. A first reason is conceptual and reductionist. This simplified case has historically been considered as expressing the basic puzzle of hydrophobic effects. Solving this basic puzzle enables specific cases to be described by combination of what is understood for the simpler cases. A second reason is that hard-core models of solute–water interactions are expected to have direct applicability to cases of nonmacromolecular hydrophobic solutes, and those small-molecule applications can be expected to be less sensitive to specifics of the actual interactions.

Testing physical ideas of hydrophobic effects

The idea of constructing an information theory description of cavity formation in water (Hummer *et al.*, 1996) reinvigorated the molecular theory of hydrophobic effects (Hummer *et al.*, 1998*a*; Pratt, (1998); Hummer *et al.*, 2000; Pratt, 2002; Paulaitis and Pratt, 2002; Pratt and Pohorille, 2002; Ashbaugh and Pratt, 2004). One advantage of this approach is that physical hypotheses can be expressed simply in a default model. Given a fixed amount of specific experimental or simulation information, i.e., data, the quality of the predictions gives an assessment of the physical ideas that are embodied in the underlying default model. Relevant physical ideas include: whether a direct description of dense fluid packings significantly improves the predictions, or whether incorporation of de-wetting of hydrophobic surfaces is required, or whether specific expression of the roughly tetrahedral coordination of water molecules in liquid water is the most helpful next step for these theories.

The information theory approach studied here grew out of earlier studies of formation of atomic sized cavities in molecular liquids (Pohorille and Pratt, 1990; Pratt and Pohorille, 1992; 1993). Since we deal with rigid and spherical solutes in the discussion we will drop the explicit indication of conformational coordinates and discuss $p(n) = p_\alpha(n|\mathcal{R}^n)$. We emphasize that the overall distribution $p(n)$ is well described by the information theory with the first two moments, $\langle n \rangle_0$ and $\langle n(n-1)/2 \rangle_0$. It is the prediction of the extreme member $p(0)$ that makes the differences in these default models significant. Computing thermodynamic properties demands more than merely observing typical behavior.

To begin, we note (see Fig. 8.3) that use of the flat or Poisson default models accurately predicts the hydrophobic hydration free energy at the two-moment level, partly due to the nonmonotonicity of the convergence with increasing moment information.

Packing

A first idea is that the default model should contain a direct description of dense fluid packings that are central to the theory of liquids; see Section 4.1, p. 61.

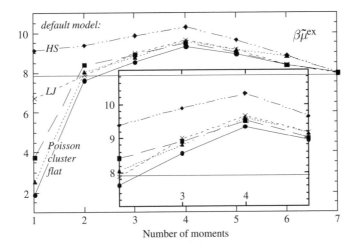

Figure 8.3 Convergence with number of binomial moments of hydration free energy $\beta\tilde{\mu}^{ex}$ predicted using several default models for a spherical solute with distance of closest approach 3.0 Å for water oxygen atoms. Identifications are: diamonds (dash–dot lines), hard-sphere default (*HS*); crosses (short dash line), Lennard–Jones (*LJ*) default; squares (long dash line), *Poisson* default; triangles (dotted line), *cluster* Poisson default; and circles (solid line), *flat* default. For this circumstance, *j*th-order binomial moments are non-zero through $j = 9$, and the horizontal line is the prediction with all nine moments included. Among the predictions at $j = 2$, the best default model is the Lennard–Jones case. But with the hard-sphere model excepted, the differences are slight. See Hummer *et al.* (1996), Gomez *et al.* (1999) and Pratt *et al.* (1999) for details of the calculations.

Accordingly, Gomez *et al.* (1999) computed $p(n)$ for the fluid of hard spheres of diameter $d = 2.67$ Å at a density $\rho d^3 = 0.633$, and adopted those results as $\hat{p}(n)$. Predictions obtained with this default model are shown in Fig. 8.3. Direct convergence is only seen if four or more moments are included. Though the convergence is more nearly monotonic from the beginning, the prediction obtained from a two-moment model is worse than from the flat and the Poisson default cases.

Attractive interactions among solvent molecules

Attractive forces between solvent molecules might play a significant role in hydrophobicity, particularly because attractive forces lower the pressure of the solvent. Dehydration of hydrophobic surfaces becomes a principal consideration for solutes larger in size than the solvent molecules (Stillinger, 1973). But perhaps such effects are being felt already for atomic solutes. Accordingly, we computed $p(n)$ for the Lennard–Jones liquid studied by Pratt and Pohorille (1992) for which attractive interactions were adjusted so that the macroscopic pressure of the solvent would be approximately zero, and adopted those results as $\hat{p}(n)$. The results of

Fig. 8.3 confirm that the results are better than for the hard-sphere default model, but also show that the convergence with number of moments is again nonmonotonic, not better than for the flat and the Poisson default models. Again, direct, nonmonotonic convergence is only seen after four occupancy moments are included.

Tetrahedral coordination of solvent molecules

The final idea checked here is whether incorporating a tetrahedral coordination structure for water molecules in liquid water significantly improves the prediction of cavity formation free energy. We used a cluster Poisson model to accomplish this (Neyman and Scott, 1956). The physical picture is of tetrahedral clusters of water molecules with prescribed intra-cluster correlations but random positions and orientations as suggested in Fig. 8.4.

This default model can be evaluated compactly (Gomez *et al.*, 1999). We assume the clusters to be tetrahedra with the oxygen atom of a water molecule at the center and at each vertex. Thus we prescribe the number of clusters to be $\rho v/5$, with v the volume of the augmented region and ρ the molecular density of the liquid water. The OO intra-cluster near-neighbor distance, the distance of a point of a tetrahedron from its center, is $2.67\,\text{Å}$ and the augmented volume is a sphere with radius $\lambda + 2.67\,\text{Å}$.

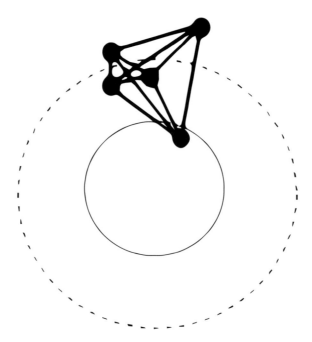

Figure 8.4 A tetrahedral cluster, the observation volume (sphere with solid outline), and the augmented volume (sphere with dashed outline). The cluster may contribute occupants of the observation volume only if the center is within the augmented volume (Gomez *et al.*, 1999).

Figure 8.3 shows the predictions for cavity formation free energy obtained with the cluster Poisson default model. The nonmonotonic convergence is still evident. The prediction utilizing two moments is more accurate than that utilizing the Poisson default model and is similar to the predictions made by the flat default or the Lennard–Jones default models.

Discussion

There are several conclusions to be drawn from this worked example. The broadest conclusion is that these PDT-based approaches provide flexibility for interrogation of physical issues underlying models of solution free energies.

Proceeding toward conclusions of higher specificity, notice that all of these applications achieve monotonic convergence only when at least four moments are utilized. The reason for this is a general one connected to the structure of the formula Eq. (4.32), p. 76; see also Fig. 4.3, p. 76. Here the most probable occupancies are $n = 3$ and 4. Other occupancies are improbable relative to those cases, and terms of Eq. (4.32) other than $n = 3$ and 4 are extremely small. But $n^{\underline{k}}$ is zero for terms $n < k$. Thus, final adjustments of the predicted probable populations await the moment information $k > 3$, which makes direct adjustments to the largest terms of Eq. (4.32). After $k > \langle n \rangle_0$, subsequent adjustments are indirect, either through the consistency and normalization requirements on the ζ_k, or through the extremely small terms of the sum.

The next general conclusion for this discussion is associated with the fact that the predictions shown in Fig. 8.3 of the different models with $k = 2$ moments cluster into two groups, one group being the hard-sphere model only. Since prediction based upon $k = 2$ moments can be qualitatively considered a shift and scaling of the default model, using $k = 2$ moments can be qualitatively viewed as adapting the default model to the present problem. This qualitative view is precisely true for the traditional continuous normal distribution. A general observation is that unless the default model is almost perfect, these information theory methods do better with a broad and less specific default model. Specific errors in a default model have to be corrected by high-order moments, and that suggests delayed convergence. The more successful group of models considered here all have broader default distributions.

For the specific physical case, we conclude that a hard-sphere solvent provides a less successful model of hydrophobicity relative to the other models considered here. Even though that hard-sphere system might be a classic initial model in the theory of liquids, specific physical conclusions on that basis should not be accepted uncritically.

A final specific conclusion, and a provocative one, concerns the relevance of tetrahedrality of solvent structure or the *clathrate likeness* of the solvent in establishing characteristic hydrophobic effects (Ashbaugh *et al.*, 2003). The cluster Poisson model builds tetrahedral structure into the default model, but crudely. How might we build a more precise default model that captures such characteristics in an organized way? The structural features of interest here are expressed in terms of molecular angular correlations, and those angular correlations are embodied in the observed moments $k = 3, 4, \ldots$ Thus one well-organized way to produce a tetrahedral default model is to use the *output* of a calculation that produces Fig. 8.3, with a specific $k = 4$, as a proposed *practically perfect* default model for this application. Figure 8.5 shows what to expect from such a nice default model: poorer predictions at $k = 2$ for just the reasons of nonmonotonic convergence that we have discussed in detail. This *goodness-of-default model* criterion therefore does not support the concept that tetrahedrality or *clathrate-likeness* is a required ingredient in theories of hydrophobic hydration (Ashbaugh *et al.*, 2003). The suggestion is that the distinctive, but weak, orientational structure associated with simulation of hydrophobic species in water is an ancillary observation, not a theory of hydrophobicity.

Note that the poorest performing default model of Fig. 8.3 – the HS default model – gives results qualitatively similar to the practically perfect tetrahedral default model, Fig. 8.5; specifically they both comfortably give nearly monotonic convergence. The spectacular accuracy of the other results of Fig. 8.3 was not anticipated, and was understood after the fact as due to an inadvertent cancellation of errors (Pratt, 2002).

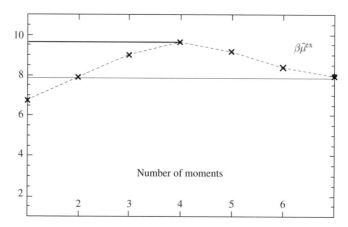

Figure 8.5 Performance of the *practically perfect* tetrahedral default model discussed above in predicting $\beta\tilde{\mu}^{\text{ex}} = -\ln p\,(0)$. Since this default model was built using the best default model of Fig. 8.3 *plus* the observed first four binomial moments, the predictions are unchanged at the $k = 4$ moment value until more than four moments are provided. The result is poorer predictions at the $k = 2$ moment level. The crosses connected by the dashed line segments are the LJ default model results of Fig. 8.3.

Entropy convergence

The results surveyed in Fig. 8.3, p. 183, prove that the two-moment information model provides a robust, physically valid description of those primitive hydrophobic hydration free energies, with the additional observation that highly specific default models, particularly the hard-sphere default model, are less successful for that purpose.

We therefore bring these models to bear on the most puzzling feature of hydrophobic effects, the temperature effects exemplified by *entropy convergence* behavior (Garde *et al.*, 1996). The convergence of entropies of hydration of hydrocarbons to a value near zero at a common temperature of approximately 385 K is well known (Baldwin, 1986; Baldwin and Muller, 1992; Fu and Freire, 1992; Lee, 1991). Entropy convergence displays a behavior common to hydrophobic effects, which become increasingly strong with increasing temperatures for temperatures below the entropy convergence temperature. Entropy convergence displays the further curiosity that the upper-temperature limit for this increasing strength with temperature is common to many cases of experimental relevance.

Figure 8.6 shows the modelled values of $\tilde{\mu}^{\mathrm{ex}}$ for spherical solutes as a function of temperature along the saturation curve of liquid water, and compares them to the chemical potentials computed directly. The quantitative agreement between the two methods is excellent over the entire temperature range. The chemical potential increases with temperature past 400 K but eventually decreases. The maximum in chemical potential occurs at about the same temperature in each case. These curves have the same shape as the experimental ones (Harvey *et al.*, 1991) for inert gases dissolved in water, but they are shifted upward due to the use of a hard-sphere model.

Entropies calculated as the temperature derivative of μ^{ex} along the saturation curve are shown in Fig. 8.7. As expected, these entropies are large and negative at room temperature, and increase with temperature. The entropies of hydration for these solutes converge in a temperature region around 400 K, close to the temperature at which they are zero. The observed entropy convergence for transfer of simple nonpolar species from the dilute gas to water (Harvey *et al.*, 1991) is similar.

A simplification of the two-moment model is obtained from the gaussian estimate

$$p(n) \approx \frac{e^{-(n-\langle n \rangle_0)^2/2\sigma^2}}{\sqrt{2\pi\sigma^2}}, \tag{8.21}$$

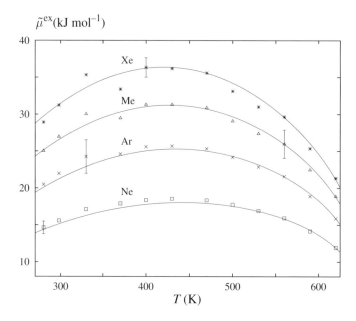

Figure 8.6 Excess chemical potentials $\tilde{\mu}^{\text{ex}}$ of model hard-sphere solutes of sizes roughly comparable to Ne, Ar, methane (Me), and Xe as a function of temperature. The hard-sphere diameters used were 2.8 Å, 3.1 Å, 3.3 Å, and 3.45 Å, respectively. The lines indicate the information theory model results and the symbols are the values computed directly with typical error bars (Garde *et al.*, 1996).

where $\sigma^2 = \left\langle (n - \langle n \rangle_0)^2 \right\rangle_0$. Remembering that $\langle n \rangle_0 = \rho v$, this gives the explicit expression

$$\mu^{\text{ex}} = -kT \ln p(0) \approx T\rho^2 \left[\frac{kv^2}{2\sigma^2} \right] + T \left[\frac{k}{2} \ln \left(2\pi\sigma^2 \right) \right], \qquad (8.22)$$

connecting the chemical potential to the density and density fluctuations of liquid water. This gaussian formula is consistent with the historical Pratt–Chandler theory (Pratt and Chandler, 1977; Hummer *et al.*, 1996, Berne, 1996; Chandler, 1993; Pratt, 2002).

The second term of Eq. (8.22) is smaller than the first, and is only logarithmically sensitive to the size of the solute. Equation (8.22) therefore says physically that the hydration free energy may be lowered by decreasing the density or the temperature of the solvent, the $T\rho^2$ factor, or by enhancing the ability of the solvent to open cavities of a size necessary to accommodate the solute, the σ^2 factor in the first term.

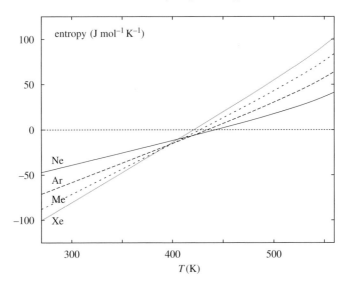

Figure 8.7 $-(\partial\mu^{\text{ex}}/\partial T)_{\text{sat}}$ along the saturation curve of liquid water for model hard-sphere solutes of sizes comparable to Ne, Ar, methane (Me), and Xe as a function of temperature. Additional equation of state contributions to the standard hydration entropy are negligible: $|(\partial\mu^{\text{ex}}/\partial T)_{\text{p}} - (\partial\mu^{\text{ex}}/\partial T)_{\text{sat}}| < 1$ and < 10 J mol^{-1} K^{-1} for temperatures $T < 450$ K and < 550 K, respectively (Garde *et al.*, 1996).

Surprisingly, $\sigma^2(T, v)$ has a negligible dependence on the temperature over that range, so that

$$\mu^{\text{ex}} \approx T\rho(T)^2 x(v) + Ty(v). \tag{8.23}$$

The quantities $x(v)$ and $y(v)$, defined by the correspondence between Eqs. (8.22) and (8.23), are only weakly dependent on the temperature. Along the saturation curve in Fig. 8.6, the combination $T\rho^2(T)$ exhibits a nonmonotonic temperature dependence. Figure 8.8 discusses schematically how this approximate formula then works. Valid guidance can be obtained by the crude estimate

$$\left(\frac{\partial\mu^{\text{ex}}}{\partial T}\right)_{\text{sat}} \propto \rho(T)^2 + 2T\rho(T)\left(\frac{\partial\rho(T)}{\partial T}\right)_{\text{sat}}, \tag{8.24}$$

which yields for the entropy convergence temperature $T_c \approx 1/(2\alpha_{\text{sat}})$, with α_{sat} the coefficient of thermal expansion along the saturation curve.

Entropy convergence and solute size

For hard-sphere solutes this entropy convergence point has a nontrivial size dependence that isn't apparent from Fig. 8.7 (Huang and Chandler, 2000; Ashbaugh and Pratt, 2004). Figure 8.9 gives a current estimate of those entropy

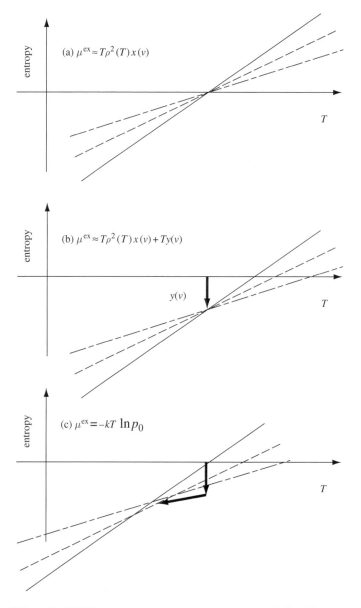

Figure 8.8 $-\left(\partial\mu^{\mathrm{ex}}/\partial T\right)_{\mathrm{sat}}$ along the saturation curve of liquid water, as in Fig. 8.7 but schematically. (a) Contribution to the entropy from the $T\rho^2(T)x(v)$ term of Eq. (8.23). This contribution dominates Eq. (8.23). (b) Sum of the contributions from both the terms in Eq. (8.23). The lack of dependence of σ^2 on temperature is assumed. (c) Entropies calculated from Eq. (4.25), p. 73, accounting for the temperature dependence of σ^2 (Garde *et al.*, 1996).

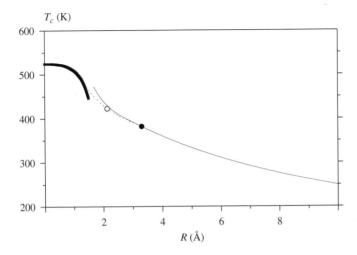

Figure 8.9 Variation of the entropy convergence temperature with increasing hard-sphere radius. The thin solid line is the convergence temperature determined under the assumption that the heat capacity is independent of temperature, and the thick solid line is the exact entropy convergence temperature for spheres smaller than $R < \sigma_{WW}/2$ (Ashbaugh and Pratt, 2004). The dashed line smoothly interpolates between the exact and constant heat capacity curves at 1.25 Å and 3.3 Å, respectively. The filled circle indicates the entropy convergence temperature of a methane-sized solute ($T_c = 382\,K$). The open circle indicates the entropy convergence temperature based on the information model ($T_c = 420\,K$) (Ashbaugh and Pratt, 2004).

convergence temperatures as a function of radius R. With increasing solute size T_c decreases so that above a radius of $\approx 8\,Å$ it is less than the normal freezing point of water. At the intermediate methane radius of 3.3 Å, however, the convergence temperature is 382 K, in excellent agreement with the experimental convergence temperature of 385 K for simple nonpolar gases and linear alkanes.

The relevance of this specific entropy convergence behavior to protein folding free energetics is problematic (Robertson and Murphy, 1997). Proteins are complicated molecules participating in both hydrophobic and hydrophilic interactions with the solution. The widely appreciated point that protein folding thermodynamics may be primarily sensitive to hydration of unfolded configurations is just as important (Paulaitis and Pratt, 2002; Pratt and Pohorille, 2002). Considering unfolded possibilities, the sizes of the obvious hydrophobic units are in the range of small-molecule hydrocarbon solutes. The largest hydrophobic side chain, that of phenylalanine, is an example. Solution thermodynamic data are available for hydrophobic solutes of just this size, e.g. for benzene, toluene, and ethyl benzene (Privalov and Gill, 1989), and those data suggest that these solutes exhibit conventional entropy convergence behavior. Thus, it is a

plausible hypothesis that entropy convergence will be expressed in protein folding thermodynamics primarily through contributions associated with the unfolded
configurations.

Pragmatic interpretation: "What are we to tell students?"

The views suggested by the model above are heretical (Dill, 1990). But that
model is sufficiently basic, successful, and compelling to require discussion
of the question, "What are we to tell students?" (Pratt and Pohorille, 2002;
Ashbaugh *et al.*, 2003).

In preparation we can note that hydrophobicity as judged by hydration free
energy is greatest at moderately elevated temperatures >100 °C along the vapor
saturation curve, as was emphasized by Murphy *et al.* (1990). Furthermore, the
most provoking puzzle for molecular mechanisms of hydrophobic phenomena has
always been the apparent increase in attractive strength of hydrophobic effects
with increasing temperature for temperatures not too high. This point is experimentally clear in the phenomenon of cold denaturation wherein unfolded soluble
proteins fold *upon heating* (Franks and Hatley, 1985; Hatley and Franks, 1986;
Franks *et al.*, 1988; Privalov, 1990). Thus it is important to explain hydrophobicity at temperatures as high as, or even higher than, conventional physiological
conditions. In contrast, molecular structural pictures such as clathrate models – as
discussed in Ashbaugh *et al.* (2003), Laviolette *et al.* (2003), and Fig. 8.10 – seem
to point to low temperature regimes and behaviors as identifying the essence of
hydrophobicity.

Two characteristics of liquid water that are relevant to the present answer are
the equation of state characteristics shown in Figs. 8.11 and 8.12. The isothermal
compressibilities shown in Fig. 8.11 indicate that water is stiffer than organic
solvents, and that the stiffness is only weakly temperature-dependent. We don't
propose here a detailed explanation of that stiffness – it is due to intermolecular
interactions among solvent molecules, hydrogen bonding in the case of liquid
water (Debenedetti, 2003) – but the present empirical theory of hydrophobicity
merely exploits those results. This stiffness is the principal determinant of the
low solubility of inert gases in liquid water. In the simplest information models
this stiffness, and its temperature dependence, is expressed by the experimental
$\langle n(n-1)\rangle_0$, which is distinctive of liquid water.

The second characteristic in our answer is the variation of the liquid density
along the liquid–vapor coexistence curve in the temperature regimes of interest
here. The coefficient of thermal expansion along the coexistence curve, plotted in
Fig. 8.12 for several solvents, is typically more than five times smaller for water
than for common organic solvents. It is a secondary curiosity that liquid water has

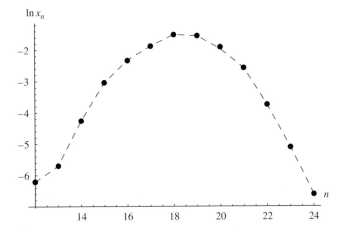

Figure 8.10 The distribution of the number of oxygen atoms within 5.1 Å of the Kr atom in aqueous solution at an elevated temperature in the region of the entropy convergence temperature (LaViolette *et al.*, 2003). These results were obtained to investigate the possibilities of clathrate nucleation upon quenching; see Filipponi *et al.* (1997) and Bowron *et al.* (1998). Note that the coordination numbers $n = 20$ or $n = 24$, which are associated with clathrate cages, are unexceptional in this distribution for the liquid solution. The subtle structure in this distribution for n below the mode may be reflective of possibilities for alternative thermodynamic phases, e.g. the coexisting gas phase, or structures with commodious cages.

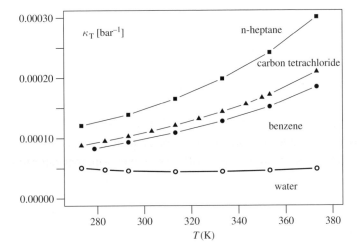

Figure 8.11 Isothermal compressibilities, $\kappa_T = -(1/\mathcal{V})(\partial\mathcal{V}/\partial p)_T$, for several solvents plotted as a function of temperature along their saturation curves. The results are taken from Rowlinson and Swinton (1982). Liquid water is stiffer than the other solvents here, and that stiffness is less temperature-dependent.

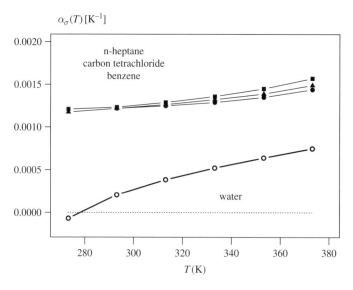

Figure 8.12 Coefficient of thermal expansion, $\alpha_{\text{sat}}(T) = 1/\mathcal{V}\,(\partial\mathcal{V}/\partial T)_{\text{sat}}$, along the liquid–gas saturation curve as a function of temperature for several solvents. The bottom curve is liquid water. The top band are the organic solvents CCl_4, benzene, and n-heptane. Of course, this quantity is negative for liquid water below $4\,°C$, and that region isn't plotted here. Water expands more slowly than the other solvents shown here. All results are from Rowlinson and Swinton (1982).

a small regime of density increase with increasing temperature; we are interested here in a much broader temperature region. Also, the critical temperature of liquid water is significantly higher than is the case for the comparative organic solvents. In consequence of these observations, the densities of typical organic solvents decrease more rapidly with increasing temperature than does the density of water in the temperature region of interest here.

These two points lead to a picture in which the aqueous medium is stiffer over a substantial temperature range, and expands with temperature less significantly than the natural comparative solvents. If these structural features of the aqueous medium are thus tempered relative to normal changes with increasing temperatures, then at higher temperatures the solvent exerts a higher kinetic pressure through collisions with hydrophobic solutes which experience principally repulsive forces on those encounters. These collisions are more energetic proportionally with increasing temperature – this is the leftmost factor of T in Eq. (8.22), p. 188. The aqueous environment thus becomes more unfavorable for hydrophobic solutes with increasing temperature. The rate of density decrease with increasing temperature eventually dominates this mechanism at the highest temperatures of interest here, and less unconventional behavior is then expected. This paraphrase of the model above is a realistic response to the question, "What are we to tell students?"

Hydrogen bonding, tetrahedral coordination, random networks and related concepts are not direct features of this answer. Nevertheless, they are relevant to understanding liquid water; they are elements in the bag of tricks that is used to achieve the engineering consequences that are discussed in the picture above (Pratt and Pohorille, 2002; Ashbaugh *et al.*, 2003).

This argument has descended onto the equation of state as a principal determinant of peculiar temperature dependences of hydrophobic effects. The statistical thermodynamic model discussed above, however, started with probabilities and fluctuations. But equations of state and fluctuations are connected by the most basic of results of Gibbsian statistical mechanics, e.g. Eq. (2.24), p. 27. Ad hoc models, such as the more simplistic lattice gas models, can be adjusted to agree with solubility at a thermodynamic state point, but if they don't agree with the equation of state of liquid water more broadly they can't be expected to describe molecular fluctuations of liquid water consistently and realistically. Thus, models of that sort are unlikely to be consistent with the picture explored here.

A broader view

It is difficult to overstate the breadth of scientific and technological interest in liquid water. The breadth of interest leads to a wide range of mechanistic speculations about molecular-scale phenomena of aqueous solutions. As a recent example, we note disparate molecular statistical thermodynamic theories of micellization (Bock and Gubbins, 2004; Maibaum *et al.*, 2004). The present answer to, "What are we to tell students?" was designed aggressively to limit mechanistic speculations. Once that response is appreciated, however, it suggests a broader view of water as the *matrix of life* (Ball, 1999; Franks, 2000). Specifically, this broader view is that liquid water, compared with other possibilities, widens the temperature domain over which biomolecular structures are stable and functional.

This is the view of hydrophobic effects suggested here. But satisfactory hydration of polar and ionic moieties is also essential for practical cases of soluble biomolecules. For hydrophilic hydration the simplest relevant observation about water as a medium is that it has a high dielectric constant. We know from preceding considerations – see Section 4.2, p. 67, for example – that this high dielectric constant is correlated with good solubility of charged and polar molecules. Further, we know that a dielectric constant as high as that for liquid water also serves to make interactions of charged and polar groups less sensitive to other variations of the solvent properties.

It is also important to recognize that water molecules participate in aqueous phase chemical processes, specifically acid–base chemistry. Chemical buffering that is universal in biophysical systems serves to mitigate changes in charge state of macromolecules. In this respect, also, liquid water moderates the possibilities for catastrophes associated with changes in solvent properties.

The natural view of these arguments is, therefore, that biochemical processes have adapted to occupy a medium offering reduced sensitivity to alterations of solvent qualities. In comparison to media and processes with higher sensitivity to changes of solvent quality, such an adaptation should be advantageous.

Coda

It is important that the theories and discussions above have considered simple models – hard spheres – of small-molecule hydrophobic solutes. As noted above, this is partly because that problem historically has been regarded as a basic puzzle of hydrophobic phenomena. The tools developed in this book have provided a compelling analysis of that basic puzzle.

But the *size* dependence of these results has long been an interest also (Stillinger, 1973). For hard-sphere model solutes, the size dependence is analyzed by introducing

$$\rho G(\lambda)4\pi\lambda^2 d\lambda = -\left(\frac{\partial \ln p_\lambda(0)}{\partial \lambda}\right) d\lambda. \tag{8.25}$$

Following the notation of Section 7.6, specifically p. 160, λ is the distance of closest approach of the solvent (water) center to the hard spherical solute. The left side of Eq. (8.25) is the differential work done in expanding the solute sphere against the solvent pressure. $G(\lambda)$, introduced on p. 121, Eq. (5.70), is the contact value of the radial distribution function of solvent centers from the position of a hard-spherical solute. $G(\lambda)$ then gives molecular-scale structural information to obtain that solvent pressure, and Fig. 8.13 shows the current best information on that molecular-scale pressure (Ashbaugh and Pratt, 2004).

We don't pursue a further detailed discussion of these results here, but confine ourselves to a few broad observations. First, the theories and discussions above have focused on hard-spherical solutes of size located roughly by the maximum of $G(\lambda)$, Fig. 8.13. These solutes are candidates for *most hydrophobic* because the solvent pressure is greatest for those sizes. The location of that maximum gives a convenient size to discriminate between small and large molecule scales for these hydration problems.

The fact that the maximal $G(\lambda)$ is substantially larger than one indicates that the local density contacting the solute surface is relatively large. In that case,

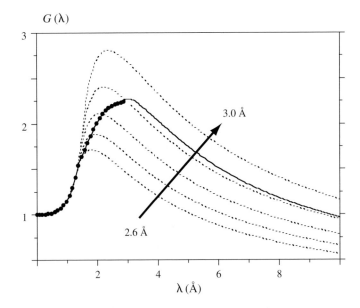

Figure 8.13 The function $G(\lambda)$ for water at 300 K at the liquid saturation conditions (Ashbaugh and Pratt, 2004). The points are obtained by direct Monte Carlo calculation, and the solid line by matching an empirical thermodynamic model for the large solute case. The dashed lines are the classic scaled-particle model (Pierotti, 1976) predictions for several solvent hard-sphere diameter parameters between $\sigma_{\mathrm{WW}} = 2.6\,\text{Å}$ and $3.0\,\text{Å}$ in $0.1\,\text{Å}$ increments. Notice that the parameter value that provides the best fit of the classic scaled-particle model for small radii is not the same as that for the large radii results.

the physical discussion of the van der Waals models, Section 4.1, p. 61, has a chance of being physically valid. The most important physical observation about the behavior of $G(\lambda)$ for large radii is that the physical problems for those length scales are likely to be sensitive to attractive solute–solvent interactions (Weeks, 2002; Zhou *et al.*, 2004). This sensitivity can be appreciated by noting that in those cases where the $G(\lambda)$ is substantially less than one, the predicted contact density isn't high, and the van der Waals argument of Section 4.1 probably doesn't apply. This is our second broad observation.

Our final observation focuses on the length scales involved in results such as shown in Fig. 8.13. We have noted the length $\lambda \approx 3\,\text{Å}$ that locates the maximum. Further detailed analyses permit extraction of additional physically interesting length scales, e.g., characterizing curvature effects (Ashbaugh and Pratt, 2004). But a nice element of context is obtained by noting a classic identification of Egelstaff and Widom (1970) of a correlation length for liquids viewed broadly. Results comparing water with several organic solvents are shown in Fig. 8.14. The

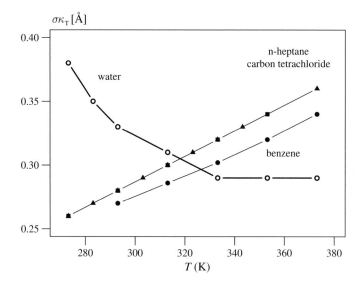

Figure 8.14 The product of the tension of the liquid–vapor surface, σ, and the isothermal compressibility, κ_T, identified by Egelstaff and Widom (1970) as proportional to the spatial correlation length. This combination was suggested as appropriate for the low density of the coexisting vapor phase.

magnitudes of those lengths are as expected (Egelstaff and Widom, 1970), but the temperature variations are distinctly different, and it is the temperature variations that are the issues of significance for hydrophobic effects. This reinforces the point that temperature dependences intrinsic to liquid water can be peculiar, and this point has to be remembered when the additional issue of hydrophobicity is considered.

8.3 Primitive hydrophilic phenomena: ion hydration

The discussion of the previous section emphasized that hydrophobic effects, as peculiar as they are, are only one side of the coin for aqueous solution chemistry and for the biophysics and biochemistry that inevitably involves water. If we are interested in molecules in aqueous solution, those molecules are likely to be hydrophilic to some extent.

Atomic ions, such as $Li^+(aq)$, are common and chemically the simplest hydrophilic aqueous solution species. The hydration free energies are known to be in the range of $-100\,kcal\,mol^{-1}$ (Friedman and Krishnan, 1973). This magnitude is comparable to chemical bond energies, and therefore the molecular-scale details of ion hydration are typically idiosyncratic in ways that are characteristic of chemical effects. The example of Section 7.3, p. 149, made it clear that chemistry is the issue of first importance for the Be^{2+}(aq) ion. Figure 8.15 compares primitive aspects of the hydration structure of the simplest comparable ions in aqueous solution, Li^+, Na^+, and K^+ (Asthagiri *et al.*, 2004c).

This simplest comparison is interesting partly because of the famous observation of Friedman and Krishnan (1973, see Table III) that the sum of standard hydration entropies for K^+ and Cl^- is about twice the standard hydration entropy of Ar, with 2 Ar having an entropy loss on dissolution about 20% larger than in the case of KCl. For methanol as a solvent, the situation is different: these solution entropies are different by nearly a factor of three, and the entropy loss on dissolution is higher for the ions. This observation is a severe challenge for mechanistic interpretations of hydration entropies.

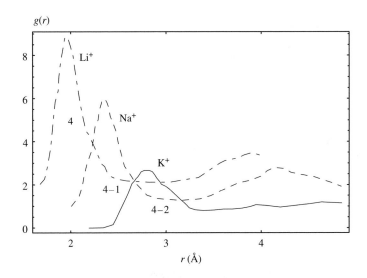

Figure 8.15 Comparison of radial distributions of oxygen atoms conditional on the simplest metal ions in typical aqueous solutions obtained by *ab initio* molecular dynamics (AIMD). See Asthagiri *et al.* (2004c) for details. The potassium result was presented by itself in higher detail in Fig. 7.7, p. 157. Notice that the lithium result (displaced vertically by 2) and the sodium result (displaced vertically by 1) have inner shells clearly defined on the basis of the $g(r)$. For lithium, the occupancy of that inner shell is almost exclusively 4. For sodium, the principal occupancy is 4, but there is a statistical admixture of another oxygen that also serves to blur the primary minimum; this occupancy is indicated by $4-1$. For potassium, this statistical characterization is $4-2$, as was also shown differently by Fig. 7.7; this leads to the occultation of the principal minimum in that case.

Nevertheless, many ionic solutes in water display standard hydrophobic behavior such as closed loop coexistence curves (Xu *et al.*, 1991; Weingartner and Steinle, 1992). So the possibilities for ion hydration range from hydrophobic to primitive chemical interactions, and leave lots of room for molecular-scale complexity in between.

Aqueous solutions offer yet another complexity because typically the dissociation

$$H_2O \rightleftharpoons HO^- + H^+$$

is significant in aqueous-phase chemistry and biochemistry. The ions HO^-(aq) and H^+(aq) are *intrinsic* to aqueous materials, and molecular-scale processes in aqueous solutions typically require buffering to control the levels of these ions. This reaction is the most fundamental among all chemical reactions in aqueous solution chemistry (Stillinger, 1978). Since this dissociation is a chemical process, it is reasonable to expect that a chemical description is necessary for the hydration of these ions.

The discussion that follows treats the hydration of each of these ions in turn. These are current research problems, and current research opinion is nonuniform. At the least, quasi-chemical treatments can serve as initial physical theories – sanity checks – and subsequent refinements can be seen in the light of a primitive physical theory. For the purposes of this book, the focus on these important ions is justified by the fact that they are likely to be tricky cases, and this permits us to investigate quasi-chemical theories in situations that make different demands on the theory than the simplest metal ions do.

HO⁻(aq)

We consider the inner-shell reactions

$$HO^- + nH_2O \rightleftharpoons HO[H_2O]_n^-. \tag{8.26}$$

The free energy changes for these reactions were calculated using standard electronic structure programs; the approach was that of the primitive quasi-chemical model of Section 7.3, p. 149, but the particular details are in Asthagiri *et al.* (2003*d*).

The results are summarized in Fig. 8.16, which is comparable to Fig. 7.2, p. 152. This simple theory predicts the predominance of the $HO \cdot [H_2O]_3^-$ quasi-component, and that prediction is independent of the level of sophistication of electronic structure theory; calculations with conventional basis sets give the same trends as calculations with larger basis sets (Asthagiri *et al.*, 2003*d*).

A simple rationalization of the electronic structure results for $HO \cdot [H_2O]_3^-$ and $HO \cdot [H_2O]_4^-$ is the following: the nominal hydroxide hydrogen atom in these

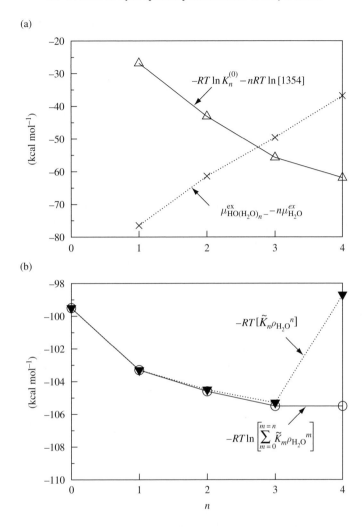

Figure 8.16 (a) Quasi-chemical contributions to the hydration free energy of HO⁻(aq) as a function of the inner-shell coordination number n. (b) Quasi-chemical estimate of the hydration free energy, with individual terms in the sum shown separately. An observation volume of radius 1.7 Å centered on the anionic oxygen defined the inner shell. Change of that radius, say to 2.0 Å, would change the $n = 0$ contribution roughly by a factor of (1.7/2.0). But that wouldn't change the net result substantially since the $n = 3$ contribution dominates, and the ion is nearly buried by the ligands in that case as was discussed in Sections 7.3 and 7.4 (Asthagiri *et al.*, 2003d).

negative ions is less positively charged than is typical of water hydrogen atoms. As a result, opportunities for hydrogen bond donation to that nominal hydroxide hydrogen are less favorable. The fourth water ligand then prefers to crowd among the other three on the oxygen side of the hydroxide anion.

Figure 8.16 shows that the chemical binding contributions for inner-shell water additions to $HO \cdot [H_2O]_3^-$ are in fact favorable. Accounting for the activity of liquid water on the basis of the ideal ρ^m factors makes formation of larger complexes even more favorable; as in Fig. 7.2, p. 152, this is the routine observation. Including hydration effects, in particular the differential hydration of ligands bound and unbound, makes the addition of water to $HO \cdot [H_2O]_3^-$ unfavorable. This emphasizes the significant role of hydration in establishing probable coordination numbers.

Another common observation in applying a quasi-chemical approach to ion hydration problems is that aggregates beyond the most probable size begin to find favorable outer-shell placements for the later additions. This seems to be the case in the present problem too. Alternative arrangements of four water molecules, such as $HO \cdot [H_2O]_3 \cdot [H_2O]^-$, are routinely found to be more favorable than $HO \cdot [H_2O]_4^-$. Numerous such arrangements are possible (Chaudhuri *et al.*, 2001). In a specific case considered by Asthagiri *et al.* (2003*d*), the fourth water molecule hydrogen bonds with the inner-shell water molecules to form a structure similar to OHW4III in Fig. 8.2 of Chaudhuri *et al.* (2001).

The lower energy of the outer-shell arrangement of the fourth water was confirmed spectroscopically by Robertson *et al.* (2003). Those experiments showed shell closure by the ligating water molecules when three water molecules are hydrogen bonded to the HO^- ion.

Critique

Tetrahedral (Mootz and Stäben, 1992) and octahedral (Rustad *et al.*, 2003) coordination environments are known for HO^- in crystalline hydrates. Evidently the numbers and arrangements of water molecules coordinating an HO^- ion are flexible enough to be decided by a crystal environment. Therefore development of the self-consistent molecular field models suggested by Section 7.8 would be valuable. Proximity of a specific cation is an issue, in general, for crystals. But it is interesting that, in crystalline NaOH hydrates beyond the monohydrate, the counter-ion is excluded from the inner hydration shell of both Na^+ and HO^- (Rustad *et al.*, 2003). The latter work used the PBE electron density functional, and found overall excellent results for crystalline NaOH hydrates. So that electron density functional model is able to properly characterize higher coordination structures where they are known to exist.

AIMD for $HO^-(aq)$

AIMD (*ab initio* molecular dynamics) calculations were carried out to follow up the nontrivial results just described for the simplest physical theory; the full details are in Asthagiri *et al.* (2004*b*). The electron density functional models that underlie current AIMD calculations are not perfect, and the AIMD work in Asthagiri *et al.* (2004*b*) investigated more than one such density functional. We discuss just one of those sets of results, the set likely to be the most satisfactory of those considered. These results utilized the rPBE (revised PBE) functional, comparable to the discussion of Section 7.4. But the broad pictures from the different functionals are qualitatively similar.

The system considered included one hydroxide anion in a periodic cube of 32 water molecules. The box size was 9.8788 Å, consistent with the experimental partial specific volume of HO^-(aq) (Marcus, 1985). All the hydrogen atoms were the deuterium isotopes, thus simulating the classical statistical mechanics of aqueous DO^- in D_2O. This system was aged through several steps, including utilization of classical model force fields and alternative electron density functional models (Asthagiri *et al.*, 2004*b*). Finally, it was aged with the rPBE functional for 5.9 ps, and a further production run of 5.9 ps was conducted. The mean temperature was 313 K. $\sqrt{\delta E^2}/|\bar{E}|$ was 2.0×10^{-5}. The drift in the relative energy was about $5 \times 10^{-6} \, ps^{-1}$.

Figure 8.17 introduces the geometric notation used in analyzing the coordination of HO^- (aq). Figure 8.18 shows the coordination number at each time, and

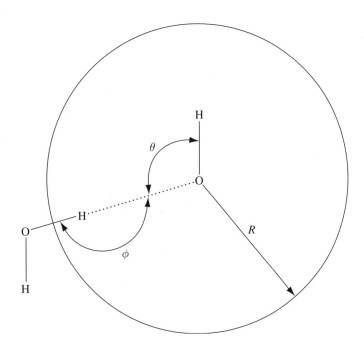

Figure 8.17 R, is the radius of the observation volume centered on the hydroxide oxygen. θ and ϕ identify the angles that specify the directionality of the hydrogen bond to water. The hydroxide hydrogen, uppermost here, is not included in the coordination number counts or in the radial distribution functions shown later.

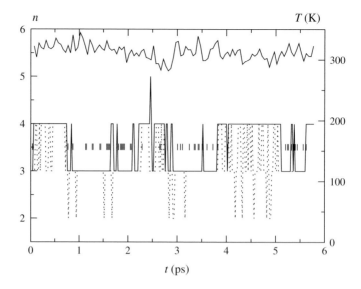

Figure 8.18 Coordination number and temperature versus time. The block-averaged temperature is shown by the solid line, and the mean temperature is 313 ± 21 K. The short vertical bars at the $n = 3.5$ level flag hydrogen exchange events, which also change the identity of the hydroxide. Note that many hydrogen exchange events occur without intercession of the $n = 4$ configuration. The dashed line applies to the selection criterion involving $R \le 2.5$, $\theta \ge 80°$, $\phi \ge 150°$; for the solid line, $R = 2.5$ Å defines the observation volume. Note that many hydrogen exchange events occur without intercession of the $n = 4$ configuration.

also the instantaneous temperature observed. Radial distribution functions are shown in Fig. 8.19. Table 8.1 presents the average fractional coordination number populations, and Table 8.2 records averages of hydrogen bonding angles using the notation of Fig. 8.17.

Consulting Fig. 8.19, it is clear that $R \le 2.5$ Å is a reasonable selection criterion, and that $n = 3$ is the prominent case. Tables 8.1 and 8.2 provide guidance on whether many of these $n = 4$ configurations should be excluded as not hydrogen-bonded.

Though the $n = 3$ case is prominent, it is clear also that the primitive quasi-chemical model seriously underestimates the population of the $n = 4$ case – see Table 8.1. Using the $R \le 2.5$ Å criterion, we find about equal populations of $n = 3$ and $n = 4$. Tightening this criterion by 0.25 Å drops the $n = 4$ population by 40% relative to the $n = 3$ case. Table 8.1 shows that a permissive $\theta \ge 80°$ cut-off excludes more of these $n = 4$ cases. The configurations thereby excluded are on the "forward" side of the hydroxide–water complex, i.e., $\theta < 90°$ in Fig. 8.17.

The distance-order decompositions of the radial distributions – Fig. 8.19 – are particularly interesting. The fourth-nearest oxygen atom builds a shoulder on the

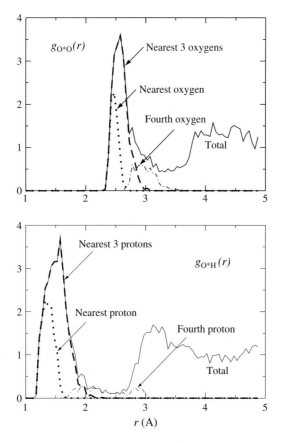

Figure 8.19 Density distribution of water oxygen and proton around the hydroxide oxygen. The distributions of the neighboring atoms are also separated into contributions according to distance order. The hydrogen of the nominal HO chemical bond, the nearest H, isn't included in this distance ordering.

outside of the principal maximum of the O*O radial distribution functions. The contributions from the nearest three protons and the nearest three oxygens are concentrated, and those protons are about 1 Å nearer the hydroxide oxygen. In contrast, the contributions from the fourth-nearest proton and fourth-nearest oxygen are diffuse and overlapping; the contribution from the fourth-nearest proton is not always inside the contribution from the fourth-nearest oxygen. Of course, those atoms are not necessarily directly bonded. These observations suggest that the fourth-nearest water molecule is not always participating in a structurally conventional hydrogen bond, but is less specifically arranged. This description as a whole is consistent with the hypothesis that the tri-coordinated species is a prominent species, though the tetra-coordinated species is also present to some extent.

Table 8.1 *Relative populations* $\hat{x}_i = x_i/x_3$ *for different selection criteria defined by Fig. 8.17. The primitive quasi-chemical theory results are denoted by PQC.*

Method	Criteria	$\hat{x}_1$	$\hat{x}_2$	$\hat{x}_3$	$\hat{x}_4$
PQC	–	0.03	0.26	1.0	0.0
AIMD	$R \leq 2.5$	–	–	1.0	1.02
AIMD	$R \leq 2.25$	–	0.01	1.0	0.60
AIMD	$R \leq 2.5, \theta \geq 80°$	–	0.01	1.0	0.43
AIMD	$R \leq 2.5, \theta \geq 80°, \phi \geq 150°$	–	0.07	1.0	0.36

Table 8.2 *Mean values of angles (degrees) defined by Fig. 8.17. The jth value pertains to the jth nearest coordinating proton after the chemically bonded H atom.*

j	1	2	3	4
$\langle \theta_j \rangle$	109.1 ± 9.9	108.0 ± 11.9	103.4 ± 13.1	91.8 ± 24.5
$\langle \phi_j \rangle$	169.8 ± 5.4	168.0 ± 6.0	165.0 ± 7.9	124.9 ± 45.8

Note also – Fig. 8.19 – that the *nearest* water-oxygen is located about 2.45 ± 0.1 Å from the hydroxide oxygen. This distance is close to the O–O separation in the calculated gas-phase structure of $HO \cdot [H_2O]^-$, 2.46 Å. We conclude that $HO \cdot [H_2O]^-$ is a prominent sub-grouping in the $HO \cdot [H_2O]_n^-$ ($n = 2, 3, 4$) species.

The $\langle \theta_j \rangle$ results of Table 8.2 document the interesting point that the three nearest coordinating protons are physically equivalent as far as averages go, and approximately disposed towards the corners of a tetrahedron. The fourth-nearest proton is distributed broadly about the plane containing the hydroxide oxygen and perpendicular to the OH chemical bond. These values are in good agreement with angles (110°) obtained from the optimized $HO \cdot [H_2O]_3^-$ cluster. $\langle \theta_4 \rangle$ is different, typically located closer to the equatorial plane, but with bigger statistical dispersion. $\langle \theta_4 \rangle$ is also different from the angle (116°) obtained from the optimized $HO \cdot [H_2O]_4^-$ cluster. The angles $\langle \phi_j \rangle$ indicate that the coordinating OH bond is not collinear with the O*O vector, and this is consistent with the cluster results. Note specifically that the water oxygen atom determining the angle ϕ_j doesn't correspond uniquely to a particular distance order for oxygen atoms; this angle is defined by the distance ordering of the hydrogen atoms, and the oxygen atoms to which those hydrogen atoms are directly bonded.

These AIMD results – Fig. 8.19 – are roughly consistent with the inferences formulated upon neutron scattering from 4.6 M NaOD aqueous solutions (Botti *et al.*, 2003) which report a mean coordination number of 3.7 ± 0.3. This value is consistent with integrals of the results of Fig. 8.19, and suggests the natural interpretation of 70% tetra-coordinated and 30% tri-coordinated species, qualitatively consistent with the view that the tri-coordinated species is substantially represented. But the results suggest the subtlety that the total population obtained by integrating to the first minimum includes a substantial fraction of the fourth-most-distant water molecules. These water molecules are not involved chemically like the nearer three water molecules. It should be additionally noted that the interpretation of the neutron scattering data involves empirical potential structure refinement (EPSR) modelling (Botti *et al.*, 2003) that produces radial distribution functions that have some interesting differences with the recent AIMD results (Tuckerman *et al.*, 2002; Zhu and Tuckerman, 2002; Chen *et al.*, 2002), including the results of Fig. 8.19. For example, with all AIMD results the maximum value of g(O*O) is less than six, and the principal peak of g(O*O) shows perceptible asymmetry on the outer side of the principal maximum, as exemplified by the identification of the contribution of the fourth-most-distant oxygen contribution in Fig. 8.19. In contrast, the EPSR model of the neutron scattering data show maximum values that exceed six, and that principal peak seems qualitatively less asymmetric. That EPSR modelled first peak is reported to occur at $\approx 2.3\,\text{Å}$, which is significantly shorter than the anticipated value $2.45\,\text{Å}$ discussed above. Note that those experiments and the present calculations differ in concentration and temperature.

The identification of $HO \cdot [H_2O]^-$ as a prominent sub-grouping agrees with spectroscopic studies on concentrated hydroxide solutions. The infrared and Raman spectra of concentrated hydroxide solutions have been interpreted in terms of $HO \cdot [H_2O]^-$ as a principal structural possibility for those systems (Zatsepina, 1971; Schiöberg and Zundel, 1973; Librovich *et al.*, 1979; Librovich and Maiorov, 1982).

This $HO \cdot [H_2O]^-$ sub-grouping also concisely resolves the high effective (not microscopic) hydration numbers extracted from dielectric dispersion measurements (Buchner *et al.*, 1999). A super-grouping of hydrated $HO \cdot [H_2O]^-$, one involving several more water molecules, could well be relevant to the time scale of the measurement, a possibility suggested by Agmon (2000).

These comparisons teach us about the performance of this simplest physical theory. An important point is how the inner shell should be defined to make reasonable statistical thermodynamic predictions. As with the $K^+(aq)$ case of Fig. 8.15, a naive eyeball analysis of a radial distribution function might not be the wisest for this assignment. On physical grounds, it has been argued that the inner-shell volume should be chosen aggressively small so that subsequent approximations such as a harmonic approximation for the optimized structure have the best chance of being valid (Pratt and Rempe, 1999). But the discussion of Section 7.4, p. 153, pointed out that this question has a variational answer – see Fig. 7.6,

p. 156. We expect it would be instructive to return to that variational perspective for both the K^+(aq) and HO^-(aq) cases.

H^+(aq)

With this example we also address the issue that quasi-chemical approaches sometimes offer flexibility in designing an inner shell, and differently designed approaches permit us to learn different features from the molecular statistical thermodynamic calculations.

As a preliminary point, we note that with a liquid material composed of n O atoms, $2n$ H atoms, and one H^+ ion, and with a specific spatial configuration of those nuclei sampled from an AIMD calculation, it isn't entirely trivial to identify one H nucleus as locating the H^+(aq) species. This basic problem also arises with the HO^- case discussed above, though the net composition is different. For the HO^-(aq) AIMD calculations a satisfactory way to identify the hydroxide ion is conceptually to associate each H nucleus with the nearest O nucleus. An alternative is to find the two nearest H nuclei to each O nucleus, and to identify the hydroxide O as the one with the largest difference between the nearest and second nearest H nucleus. Those two approaches invariably gave the same answer with those AIMD data.

Those alternatives have analogs for the H^+(aq) case too. We might assign each H nucleus to its closest O nucleus. We expect on physical grounds that such a procedure would identify one H_3O^+ ion – though that expectation is not guaranteed. Alternatively, we might find the shortest third-nearest OH distance, and regard that O nucleus as the center of an H_3O^+ ion. In the latter case, an H nucleus is distinguished in addition to an O nucleus that might be taken as the centering nucleus for an H_3O^+ ion.

But it is an interesting physical question whether an O nucleus centers the ionic structure that is the most appropriate representative of H^+(aq). The Zundel ion – see Fig. 8.20 – is another possibility, an H-centered proposal. To investigate this possibility, we choose our distinguished H^+ ion to be identified as discussed in the previous paragraph. We choose the indicator function discussed in basic fashion in Chapter 7 to require an oxygen atom within a spherical observation ball, and further require that O atom to be carrying two additional H atoms at least as close to it as the distance to the central H atom. The detailed specification might be comforting logically, but we don't expect the details to have high practical relevance, involved as they are.

With this set up, we then study equilibria for chemical equations

$$H^+ + mH_2O \rightleftharpoons H(H_2O)_m^+ \tag{8.27}$$

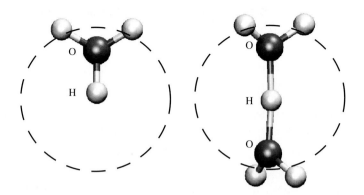

Figure 8.20 Configurations considered in the primitive quasi-chemical model of this section. The complex on the right, $H_5O_2^+$ is the Zundel structure; on the left is the classic oxonium (hydronium) ion.

describing formation of inner-shell complexes. The present formulation has the advantage that only two cases are significant, the cases shown in Fig. 8.20. The free energy changes for these reactions were calculated using standard electronic structure programs as above, but the particular details are in Grabowski *et al.* (2002) and Asthagiri *et al.* (2003*b*).

Utilizing a dielectric continuum model for the outer-shell contribution (Grabowski *et al.*, 2002), the Zundel structure was found to be the dominant representative, and a hydration free energy $\mu^{\text{ex}}_{H^+(aq)} \approx -255 \, \text{kcal mol}^{-1}$ was obtained. This is within the wide range -253 to $-265 \, \text{kcal mol}^{-1}$ of tabulated values.

In this case, outer-shell electrostatic contributions were investigated with classical molecular simulation models also (Asthagiri *et al.*, 2003*b*). The same partial charges for the solute were considered, but the van der Waals interactions were those of the standard simulation model. Again the Zundel structure was predicted to be the dominant representative, the outer-shell contributions were less deep in those cases, and net values near $-244 \, \text{kcal mol}^{-1}$ were found.

As noted above, we could have formulated this problem with an O-centered definition. In that case $OH_3 \cdot (H_2O)_3^+$, the Eigen cation, would have been the principal representative. The corresponding result of a primitive quasi-chemical calculation is $-248 \, \text{kcal mol}^{-1}$. This must be regarded as within the uncertainty of all the calculations; for example, packing effects and dispersion interactions have not been considered. But the H-centered definition permits us to provide a rough comparison between the free energetics of the Zundel case and the Eigen case, with the latter alternative regarded as a specification of the outer-shell material for the left possibility of Fig. 8.20. The physical suggestion is that the Zundel case is more populous. It has been noted, however, that this simplest physical theory might be more satisfactory for more restricted inner-shell definitions with the expectation that a harmonic approximation would be more satisfactory then. Thus, it is possible that such a simple physical theory is less satisfactory for larger complex ions. In any case, the alternative H-centered definition of this problem permitted us to address the *sub-grouping* puzzle of the previous HO⁻(aq) discussion.

AIMD for H⁺(aq)

Again, the revised PBE (rPBE) functional was used; the full details are specified in Asthagiri *et al.* (2003*a*). The system is 32 water molecules and a single H^+ in a cubic box of length 9.8432 Å. All the hydrogen atoms were replaced by deuterium atoms in the *ab initio* simulation. A total of 22.1 ps of simulation was performed, of which the first 13.5 ps was for equilibration and the last 8.6 ps was for data collection. The mean temperature in the production phase was 301 ± 20 K. The relative energy fluctuation, $\sqrt{\delta E^2}/|\bar{E}|$, was 6.5×10^{-5}. The relative drift in the energy was about -3×10^{-5} ps^{-1}.

Figure 8.22 provides the density distributions around O* and H* – see Fig. 8.21. It is clear that water oxygens around O* are not symmetrically placed. O″ is always closer to O* and at a distance of about 2.45 Å. This is very close to the O*–O separation expected in the Zundel complex.

The oxygen density distribution around H* shows many interesting features. There are two oxygen atoms coordinating the H*, and an observation volume

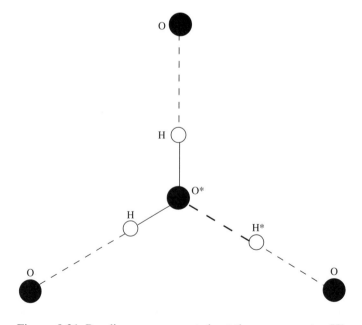

Figure 8.21 Bonding arrangement about the excess proton H*.

of radius 1.5 Å would substantially include both of them – see Fig. 8.22. But the oxygens are *asymmetrically* disposed around H*. The H*–O* bond length is shorter at about 1.14 Å. The H*–O* bond is statistically different – longer – than the 1.01 Å expected for an Eigen complex. But it is also somewhat shorter than the 1.20 Å expected for the Zundel complex.

The quasi-chemical theory suggests that, from the perspective of the proton, the Zundel cation is the most appropriate structure for describing the thermodynamics of H$^+$(aq). The simulation results support this prediction, but display additional

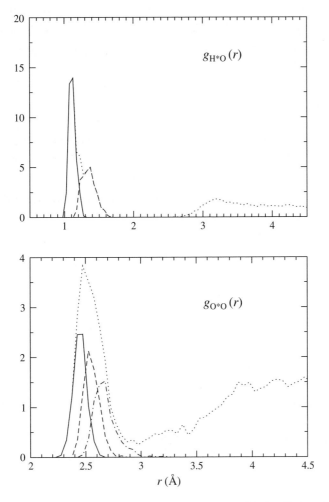

Figure 8.22 Radial distribution functions associated with the distinguished acid proton H* and its closest oxygen O*. The darker lines show decomposition according to the distance-order of the contributing atoms, i.e., the solid line is the contribution from the nearest-neighbor O atom in each case, the dark dashed curve is the contribution of the second-nearest neighbor, and so on.

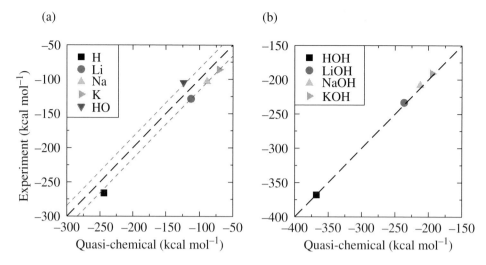

Figure 8.23 Comparison of experimental absolute hydration free energies for some monovalent ions with values calculated on the basis of the primitive quasi-chemical approximation, at ideal 1 M standard state conditions (Asthagiri *et al.*, 2003*a*), (a) with single-ion values shows an offset of positive and negative ions identifying at this level of approximation a potential of the phase contribution as discussed in Section 4.2, following p. 67. This offset vanishes with neutral combinations shown in (b).

subtlety. Most importantly, the distinguished proton is asymmetrically positioned between the two coordinating oxygen atoms.

The asymmetric placement of the proton in $H_5O_2^+$ has been noted earlier (Lobaugh and Voth, 1996; Schmitt and Voth, 1999; Vuilleumier and Borgis, 1999), and the origin traced to the classical description of proton motion. In particular, for classical nuclear motion, the central proton in $H_5O_2^+$ traverses a double-well potential which has a barrier on the order of kT for nominal oxygen–oxygen distances. Incorporating nuclear quantum effects either explicitly (Lobaugh and Voth, 1996; Schmitt and Voth, 1999) or implicitly in terms of effective potentials (Vuilleumier and Borgis, 1999) was sufficient to wash out bimodal results.

Concentrated HCl(aq) solutions have been studied by X-ray and neutron diffraction methods (Triolo and Narten, 1975; Botti *et al.*, 2004). Comparison of our results with those works is problematic because the interpretations are not uniform, and the assumptions seem to preclude some of the questions we address. For example, the interpretations assume rigid H_3O^+ molecular units.

Comparison with experimental hydration free energies

Having considered aqueous ions of both positive and negative charge types, we can make direct thermodynamic comparisons utilizing charge neutral combinations as discussed in Section 4.2, specifically on p. 69. Corrected to the ideal 1 M standard state, the combination $\mu_{H^+(aq)} + \mu_{HO^-(aq)}$ is computed to be $-244\,\text{kcal mol}^{-1} - 124\,\text{kcal mol}^{-1} = -368\,\text{kcal mol}^{-1}$ (Asthagiri *et al.*, 2003*b*); experimental values in the range $-[368, 371]\,\text{kcal/mol}$ (Asthagiri *et al.*, 2003*b*; Klots, 1981) have been offered.

A slightly broader comparison is shown in Fig. 8.23. The encouraging comparison shown should be viewed critically for several reasons. In the first place, the energies displayed are chemical scale energies, and errors that are substantial on the thermal scale kT aren't exposed. This emphasizes again that a chemical description is an essential ingredient of convincing theories of the free energies. In the second place, partial molar volumes and entropies haven't been similarly studied, and these properties are expected to test molecular-scale theories more sensitively.

References

Accardi, A. and Miller, C., Secondary active transport mediated by a prokaryotic homologue of ClC Cl⁻ channels. *Nature* **427**, 803–807 (2004).

Agmon, N., Mechanism of hydroxide mobility. *Chem. Phys. Lett.* **319**, 247–252 (2000).

Aidley, D. J. and Stanfield, P. R., *Ion Channels: Molecules in Action*. Cambridge: Cambridge University Press (1996).

Alderighi, L., Gans, P., Midollini, S., and Vacca, A., Aqueous solution chemistry of beryllium. *Adv. Inorg. Chem.* **50**, 109–172 (2000).

Allen, M. P. and Tildesley, D. J., *Computer Simulation of Liquids*. Oxford: Oxford Science (1987).

Andersen, H. C., Structure of liquids. *Annu. Rev. Phys. Chem.* **26**, 145–166 (1975). Cluster methods in equilibrium statistical mechanics of fluids. In *Statistical Mechanics. Part A: Equilibrium Techniques*, pp. 1–45, New York: Plenum (1977).

Arfken, G., *Mathematical Methods for Physicists*, 3rd edition. New York: Academic Press (1985).

Ashbaugh, H. S., Ph.D. thesis. New York: University of Delaware (1998).

Ashbaugh, H. S., Asthagiri, D., Pratt, L. R., and Rempe, S. B., Hydration of krypton and consideration of clathrate models of hydrophobic effects from the perspective of quasi-chemical theory. *Biophys. Chem.* **105**, 323–338 (2003).

Ashbaugh, H. S., Kaler, E. W., and Paulaitis, M. E., A "universal" surface area correlation for molecular hydrophobic phenomena. *J. Am. Chem. Soc.* **121**, 9243 (1999).

Ashbaugh, H. S. and Paulaitis, M. E., Effect of solute size and solute–water attractive interactions on hydration water structure around hydrophobic solutes. *J. Am. Chem. Soc.* **123**, 10 721–10 728 (2001).

Ashbaugh, H. S. and Pratt, L. R., Scaled particle theory and the length-scales involved in hydrophobic hydration of aqueous biomolecular assemblies. Technical Report LA-UR-03-2144, Los Alamos National Laboratory (2004).

Ashbaugh, H. S. and Pratt, L. R., Colloquium: Scaled particle theory and the length scales of hydrophobicity. *Rev. Mod. Phys.* **78**, 159–178 (2006).

Ashbaugh, H. S., Pratt, L. R., Paulaitis, M. E., Clohecy, J., and Beck, T. L., Deblurred observation of the molecular structure of a water–oil interface. Technical Report LA-UR-04-4750, Los Alamos National Laboratory (2004).

Ashbaugh, H. S., Pratt, L. R., Paulaitis, M. E., Clohecy, J., and Beck, T. L., Deblurred observation of the molecular structure of an oil-water interface. *J. Am. chem. Assoc.* **127**, 2808–2809 (2005).

Ashcroft, F. M., *Ion Channels and Disease*. New York: Academic Press (2000).

Ashcroft, N. W. and Mermin, N. D., *Solid State Physics*. Philadelphia: Saunders College (1976).

Asthagiri, D., Ashbaugh, H. S., and Pratt, L. R., A fresh attack on the statistical thermodynamics of molecular liquids. Technical Report LA-UR-03-5483, Los Alamos National Laboratory (2003a).

Asthagiri, D. and Pratt, L. R., Quasi-chemical study of Be^{2+} (aq) speciation. *Chem. Phys. Lett.* **371**, 613–619 (2003).

Asthagiri, D., Pratt, L. R., and Ashbaugh, H. S., Absolute hydration free energies of ions, ion–water clusters, and quasi-chemical theory. *J. Chem. Phys.* **119**, 2702–2708 (2003b).

Asthagiri, D., Pratt, L. R., and Kress, J. D., Free energy of liquid water on the basis of quasi-chemical theory and *ab initio* molecular dynamics. *Phys. Rev. E* **68**, 041505 (2003c).

An *ab initio* molecular dynamics and quasi-chemical study of H^+(aq). Technical Report LA-UR-04-1457, Los Alamos National Laboratory (2004a).

Asthagiri, D., Pratt, L. R., Kress, J. D., and Gomez, M. A., The hydration state of HO^-(aq). *Chem. Phys. Lett.* **380**, 530–535 (2003d).

HO^-(aq) hydration and mobility. *Proc. Natl. Acad. Sci. USA* **101**, 7229–7233 (2004b).

Asthagiri, D., Pratt, L. R., Paulaitis, M. E., and Rempe, S. B., Hydration structure and free energy of biomolecularly specific aqueous dications, including Zn^{2+} and first transition row metals. *J. Am. Chem. Soc.* **126**, 1285–1289 (2004c).

Badyal, Y. S., Price, D. L., Saboungi, M. L., Haeffner, D. R., and Shastri, S. D., Quantum effects on the structure of water at constant temperature and constant atomic density. *J. Chem. Phys.* **116**, 10833–10837 (2002).

Bae, Y. C., Shim, J. J., Soane, D. S., and Prausnitz, J. M., Representation of vapor–liquid and liquid–liquid equilibria for binary systems containing polymers – applicability of an extended Flory–Huggins equation. *J. App. Poly. Sci.* **47**, 1193–1206 (1993).

Baldwin, R. L., Temperature dependence of the hydrophobic interaction in protein folding. *Proc. Natl. Acad. Sci. USA* **83**, 8069–8072 (1986).

Baldwin, R. L. and Muller, N., Relation between the convergence temperatures T_h^* and T_s^* in protein unfolding. *Proc. Natl. Acad. Sci. USA* **89**, 7110–7113 (1992).

Ball, P., *H_2O: A Biography of Water*. London: Weidenfeld & Nicolson (1999).

Barker, J. A. and Henderson, D., What is liquid – understanding states of matter. *Rev. Mod. Phys.* **48**, 587–671 (1976).

Barrat, J.–L. and Hansen, J.–P., *Basic Concepts for Simple and Complex Liquids*. Cambridge: Cambridge University Press (2003).

Barthel, J., Buchner, R., Holzl, C. G., and Conway, B. E., Dynamics of molten $CF_3SO_3H \cdot H_2O$ probed by temperature dependent dielectric spectroscopy. *J. Chem. Soc. Faraday Trans.* **94**, 1953–1958 (1998).

Bates, F. S., Muthukumar, M., Wignall, G. D., and Fetters, L. J., Thermodynamics of isotopic polymer mixtures – significance of local structural symmetry. *J. Chem. Phys.* **89**, 535–544 (1988).

Beck, T. L., Quantum path integral extension of Widom's test particle method for chemical potentials with application to isotope effects on hydrogen solubilities in model solids. *J. Chem. Phys.* **96**, 7175–7177 (1992).

Real space mesh techniques in density functional theory. *Rev. Mod. Phys.* **72**, 1041–1080 (2000).

Beck, T. L., Quantum contributions to free energy changes in fluids. In A. Pohorille and C. Chipot (eds.), *Free Energy Calculation: Theory and Applications in Chemistry and Biology* (Springer-Verlag, in press, 2006).

Beck, T. L. and Klatte, S. J., Computer simulations of interphases and solute transfer in liquid and size exclusion chromatography. In J. F. Parcher and T. L. Chester (eds.), *Unified Chromatography*, pp. 67–81, Washington: American Chemical Society (2000).

Beck, T. L. and Marchioro, T. L., The quantum potential distribution theorem. In H. Grabert, A. Inomata, L. Schulman, and U. Weiss (eds.), *Path Integrals from meV to MeV: Tutzing 1992*, pp. 238–243, Singapore: World Scientific (1993).

Becke, A. D., Density-functional thermochemistry. III. The role of exact exchange. *J. Chem. Phys.* **98**, 5648 (1993).

Beglov, D. and Roux, B., Finite representation of an infinite bulk system: solvent boundary potential for computer simulations. *J. Chem. Phys.* **100**, 9050–9063 (1994).

  Dominant solvation effects from the primary shell of hydration: approximation for molecular dynamics simulations. *Biopolymers* **35**, 171–178 (1995).

Beiner, M., Fytas, G., Meier, G., and Kumar, S. K., Pressure-induced compatibility in a model polymer blend. *Phys. Rev. Lett.* **81**, 594–597 (1998).

Benmore, C. J., Tomberli, B., Egelstaff, P. A., and Neuefeind, J., Quantum effects in the structure of liquid benzene at room temperature. *Mol. Phys.* **99**, 787–794 (2001).

Bennett, C. H., Efficient estimation of free-energy differences from Monte-Carlo data. *J. Comp. Phys.* **22**, 245–268 (1976).

Berendsen, H. J. C., Grigera, J. R., and Straatsma, T. P., The missing term in effective pair potentials. *J. Phys. Chem.* **91**, 6269–6271 (1987).

Berne, B. J., Inferring the hydrophobic interaction from the properties of neat water. *Proc. Natl. Acad. Sci. USA* **93**, 8800–8803 (1996).

Bethe, H., Statistical theory of superlattices. *Proc. Roy. Soc. London A* **150**, 552 (1935).

Bizzarri, A. R. and Cannistraro, S., Molecular dynamics of water at the protein–solvent interface. *J. Phys. Chem. B* **106**, 6617–33 (2002).

Bjorling, M., Pellicane, G., and Caccamo, C., On the application of Flory–Huggins and integral equation theories to asymmetric hard sphere mixtures. *J. Chem. Phys.* **111**, 6884–6889 (1999).

Bock, H. and Gubbins, K. E., Anomalous temperature dependence of surfactant self-assembly from aqueous solution. *Phys. Rev. Lett.* **92**, 135 701–135 704 (2004).

Botti, A., Bruni, F., Imberti, S., Ricci, M. A., and Soper, A. K., Solvation of hydroxyl ions in water. *J. Chem. Phys.* **119**, 5001–5004 (2003).

  Ions in water: the microscopic structure of a concentrated HCl solution. *J. Chem. Phys.* **121**, 7840–7848 (2004).

Bowron, D. T., Filipponi, A., Roberts, M. A., and Finney, J. L., Hydrophobic hydration and the formation of a clathrate hydrate. *Phys. Rev. Lett.* **81**, 4164–4167 (1998).

Brout, R. and Carruthers, P., *Lectures on the Many-Electron Problem*. New York: Interscience Publishers (1963).

Bruno, J., Beryllium(II) hydrolysis in $3.0 \, \text{mol} \, \text{dm}^{-3}$ perchlorate. *J. Chem. Soc. Dalton Trans.* 2431–2437 (1987).

Buchner, R., Hefter, G., May, P. M., and Sipos, P., Dielectric relaxation of dilute aqueous NaOH, $NaAl(OH)_4$, and $NaB(OH)_4$. *J. Phys. Chem. B* **103**, 11 186–11 190 (1999).

Buono, G. S. D., Rossky, P. J., and Schnitker, J., Model dependence of quantum isotope effects in liquid water. *J. Chem. Phys.* **95**, 3728–3737 (1991).

Callen, H. B., *Thermodynamics and an Introduction to Thermostatistics*, 2nd edition. New York: John Wiley & Sons (1985).

Chae, D. G., Ree, F. H., and Ree, T., Radial distribution functions and equation of state of the hard-disk fluid. *J. Chem. Phys.* **50**, 1581–1589 (1969).

Chandler, D., *Introduction to Modern Statistical Mechanics*. Oxford: Oxford University Press (1987).

Gaussian field model of fluids with an application to polymeric fluids. *Phys. Rev. E* **48**, 2898 (1993).

Chandler, D. and Andersen, H. C., Optimized cluster expansions for classical fluids. II. Theory of molecular liquids. *J. Chem. Phys.* **57**, 1930 (1972).

Chandler, D., Silbey, R., and Ladanyi, B. M., New and proper integral equations for site–site equilibrium correlations in molecular fluids. *Mol. Phys.* **46**, 1335–1345 (1982).

Chandler, D., Weeks, J. D., and Andersen, H. C., Van der Waals picture of liquids, solids, and phase transformations. *Science* **220**, 787–794 (1983).

Chaudhuri, C., Wang, Y.–S., Jiang, J. C., *et al.*, Infrared spectra and isomeric structures of hydroxide ion–water clusters $OH^-(H_2O)_{1-5}$: a comparison with $H_3O^+(H_2O)_{1-5}$. *Mol. Phys.* **99**, 1161–1173 (2001).

Chen, B., Ivanov, I., Park, J. M., Parrinello, M., and Klein, M. L., Solvation structure and mobility mechanism of OH^-: a Car–Parrinello molecular dynamics investigation of alkaline solutions. *J. Phys. Chem. B* **106**, 12 006–12 016 (2002).

Chen, M. F. and Chen, T. Y., Different fast-gate regulation by external Cl^- and H^+ of the muscle-type ClC chloride channels. *J. Gen. Physiol.* **118**, 23–32 (2001).

Chen, Y.–G., Kaur, C., and Weeks, J. D., Connecting systems with short and long ranged interactions: local molecular field theory for ionic fluids. *J. Phys. Chem. B* **108**, 19 874–19 884 (2005).

Chen, Y.–G. and Weeks, J. D., Different thermodynamic pathways to the solvation free energy of a spherical cavity in a hard sphere fluid. *J. Chem. Phys.* **118**, 7944–7953 (2003).

Ciccotti, G., Frenkel, D., and McDonald, I. R., *Simulation of Liquids and Solids. Molecular Dynamics and Monte Carlo Methods in Statistical Mechanics*. Amsterdam: North-Holland (1987).

Clohecy, J., *Computer Simulation of Liquid Chromatographic Interfaces*. Ph.D. thesis. Cincinnati: University of Cincinnati (2005).

Corcelli, S. A., Kress, J. D., Pratt, L. R., and Tawa, G. J., Mixed-direct-iterative methods for boundary integral formulations of continuum dielectric solvation models. In L. Hunter and T. E. Klein (eds.), *Pacific Symposium on Biocomputing 1996*, pp. 142–159, Singapore: World Scientific (1995).

Crouzy, S., Berneche, S., and Roux, B., Extracellular blockade of K^+ channels by TEA: results from molecular dynamics simulations of the KCSA channel. *J. Gen. Physiol.* **118**, 207–217 (2001).

Cummings, P. T. and Stell, G., Interaction site models for molecular fluids. *Mol. Phys.* **46**, 383–426 (1982).

Debenedetti, P. G., Supercooled and glassy water. *J. Phys. Condensed Matter* **15**, R1669–R1726 (2003).

DePablo, J. J. and Escobedo, F. A., Monte Carlo methods for polymeric systems. *Adv. Chem. Phys.* **105**, 337–367 (1999).

Dickman, R. and Hall, C. K., Equation of state for chain molecules: continuous-space analog of Flory theory. *J. Chem. Phys.* **85**, 4108–4115 (1986).

Dill, K. A., Dominant forces in protein folding. *Biochem.* **29**, 7133–7155 (1990).

Doll, J. D., Beck, T. L., and Freeman, D. L., Equilibrium and dynamical Fourier path integral methods. *Adv. Chem. Phys.* **78**, 61–127 (1990).

Dutzler, R., Campbell, E. B., Cadene, M., Chait, B. T., and MacKinnon, R., X-ray structure of a ClC chloride channel at 3.0 Å reveals the molecular basis of anion selectivity. *Nature* **415**, 287–294 (2002).

Dutzler, R., Campbell, E. B., and MacKinnon, R., Gating the selectivity filter in ClC chloride channels. *Science* **300**, 108–112 (2003).

Efron, B. and Tibshirani, R. J., *An Introduction to the Bootstrap*. New York: Chapman & Hall (1993).

Egelstaff, P. A. and Widom, B., Liquid surface tension near the triple point. *J. Chem. Phys.* **53**, 2667–2669 (1970).

Eikerling, M., Paddison, S. J., Pratt, L. R., and Zawodzinski, T. A., Defect structure for proton transport in a triflic acid monohydrate solid. *Chem. Phys. Lett.* **368**, 108–114 (2003).

Eisenberg, D. and Kauzmann, W., *The Structure and Properties of Water*. Singapore: Oxford University Press (1969).

Escobedo, F. A. and dePablo, J. J., Chemical potential and equations of state of hard core chain molecules. *J. Chem. Phys.* **103**, 1946–1956 (1995).

Feynman, R. P., *Statistical Mechanics. A Set of Lectures*. Reading, MA: W. A. Benjamin (1972).

Feynman, R. P. and Hibbs, A. R., *Quantum Mechanics and Path Integrals*. New York: McGraw–Hill (1965).

Filipponi, A., Bowron, D. T., Lobban, C., and Finney, J. L., Structural determination of the hydrophobic hydration shell of Kr. *Phys. Rev. Letts.* **79**, 1293–1296 (1997).

Flory, P. J., Thermodynamics of polymer solutions. *Disc. Faraday Soc.* **58**, 7–29 (1970).

Fontenot, A. P., Newman, L. S., and Kotzin, B. L., Chronic beryllium disease: T cell recognition of a metal presented by HLA-DP. *Clin. Imm.* **100**, 4–14 (2001).

Fowler, R. H. and Guggenheim, E. A., Statistical thermodynamics of superlattices. *Proc. Roy. Soc. London A* **174**, 189 (1940).

Franks, F., *Water: A Matrix of Life*, 2nd edition. Cambridge: Royal Society of Chemistry (2000).

Franks, F. and Hatley, R. H. M., Low temperature unfolding of chymotrypsinogen. *CryoLetters* **6**, 171–180 (1985).

Franks, F., Hatley, R. H. M., and Friedman, H. L., The thermodynamics of protein stability. Cold destabilization as a general phenomenon. *Biophys. Chem.* **31**, 307–316 (1988).

Frenkel, D., Free-energy computation and first-order phase transitions. In *International School of Physics "Enrico Fermi,"* volume XCVII, pp. 151–188. Bologna: Soc. Italiana di Fisica (1986).

Frenkel, D. and Smit, B., *Understanding Molecular Simulation. From Algorithms to Applications*. 2nd edition. San Diego: Academic Press (2002).

Friedman, H. L. and Dale, W. D. T., Electrolyte solutions at equilibrium. In B. J. Berne (ed.), *Statistical Mechanics. Part A: Equilibrium Techniques*, pp. 85–135, New York: Plenum (1977).

Friedman, H. L. and Krishnan, C. V., Thermodynamics of ion hydration. In F. Franks (ed.), *Water: A Comprehensive Treatise*, volume 3, pp. 1–118. New York: Plenum (1973).

Fu, L. and Freire, E., On the origin of the enthalpy and entropy convergence temperatures in protein folding. *Proc. Natl. Acad. Sci. USA* **89**, 9335–9338 (1992).

Garde, S., Hummer, G., García, A. E., Paulaitis, M. E., and Pratt, L. R., Origin of entropy convergence in hydrophobic hydration and protein folding. *Phys. Rev. Lett.* **77**, 4966–4968 (1996).

De Gennes, P. G., Soft matter. *Science* **256**, 495–497 (1992).

Giaquinta, P. V. and Giunta, G., Structural remnant of the gas–liquid phase-transition in a fluid of hard-spheres. *Phys. Rev. A* **36**, 2311–2314 (1987).

Gilson, M., Given, J., Bush, B., and McCammon, J., The statistical-thermodynamic basis for computation of binding affinities: a critical review. *Biophys. J.* **72**, 1047–1069 (1997).

Gomez, M. A. and Pratt, L. R., Construction of simulation wave functions for aqueous species: D_3O^+. *J. Chem. Phys.* **109**, 8783–8789 (1998).

Gomez, M. A., Pratt, L. R., Hummer, G., and Garde, S., Molecular realism in default models for information theories of hydrophobic effects. *J. Phys. Chem. B* **103**, 3520–3523 (1999).

Grabowski, P., Riccardi, D., Gomez, M. A., Asthagiri, D., and Pratt, L. R., Quasi-chemical theory and the standard free energy of $H^+(aq)$. *J. Phys. Chem. A* **106**, 9145–9148 (2002).

Graham, R. L., Knuth, D. E., and Patashnik, O., *Concrete Mathematics: A Foundation for Computer Science*. 2nd edition. Reading: Addison-Wesley (1994).

Grossman, J. C., Schwegler, E., Draeger, E. W., Gygi, F., and Galli, G., Towards an assessment of the accuracy of density functional theory for first principles simulations of water. *J. Chem. Phys.* **120**, 300–311 (2004).

Guggenheim, E. A., The statistical mechanics of regular solutions. *Proc. R. Soc. London A* **148**, 304 (1935).

The statistical mechanics of co-operative assemblies. *Proc. Roy. Soc. London A* **169**, 134 (1938).

Guillot, B. and Guissani, Y., Hydrogen-bonding in light and heavy water under normal and extreme conditions. *Fluid Phase Equilibria* **151**, 19–32 (1998a).

Quantum effects in simulated water by the Feynman–Hibbs approach. *J. Chem. Phys.* **108**, 10 162–10 174 (1998b).

Hammersley, J. M. and Handscomb, D. C., *Monte Carlo Methods*. London: Chapman & Hall (1964).

Hammersley, J. M. and Morton, K. W., Poor man's Monte Carlo. *J. Roy. Stat. Soc. Series B – Stat. Meth.* **16**, 23–38 (1954).

Han, C. C., Bauer, B. J., Clark, J. C., *et al.*, Temperature, composition and molecular-weight dependence of the binary interaction parameter of polystyrene polyvinyl methyl-ether) blends. *Polymer* **29**, 2002–2014 (1988).

Hansen, J.–P. and McDonald, I. R., *Theory of Simple Liquids*. New York: Academic Press (1976).

Harvey, A. N., Sengers, J. M. H. L., and Tanger IV, J. C. *J. Phys. Chem.* **95**, 932–937 (1991).

Hatley, R. H. M. and Franks, F., Denaturation of lactate dehydrogenase at subzero temperatures. *CryoLetters* **7**, 226–233 (1986).

Head–Gordon, T. and Hura, G., Water structure from scattering experiments and simulations. *Chem. Rev.* **102**, 2651–2670 (2002).

Hehre, W. J., Ditchfield, R., Radom, L., and Pople, J. A., Molecular orbital theory of electronic structure of organic compounds. 5. Molecular theory of bond separation. *J. Am. Chem. Soc.* **92**, 4796–4801 (1970).

Henderson, D. and Barker, J. A., Perturbation theories. In D. Henderson (ed.), *Physical Chemistry. An Advanced Treatise*, pp. 377–412. New York: Academic Press (1971).

Hertz, J., Krogh, A., and Palmer, R. G., *Introduction to the Theory of Neural Computation*. Redwood City, California: Addison–Wesley (1991).

Hildebrand, J. H., The entropy of solution of molecules of different size. *J. Chem. Phys.* **15**, 225–228 (1947).

Hildebrand, J. H., Prauznitz, J. M., and Scott, R. L., *Regular and Related Solutions*. New York: van Nostrand Reinhold (1970).

Hill, T. L., *An Introduction to Statistical Thermodynamics*. New York: Dover (1986).

Statistical Mechanics. New York: Dover (1987).

Hille, B., *Ion Channels of Excitable Membranes*. Sunderland: Sinauer (2001).

Hirata, F., Chemical processes in solution studied by an integral equation theory of molecular liquids. *Bull. Chem. Soc. Jap.* **71**, 1483– 1499 (1998).

Hirata, F. (ed.), *Molecular Theory of Solvation*. Dordrecht: Kluwer Academic Publishers (2003).

Hoffman, G. G. and Pratt, L. R., Optimized Thomas–Fermi potential for discrete propagator electron density functional calculations. In J. D. Doll and J. E. Gubernatis (eds.), *Proceedings of the International Workshop on Quantum Simulation of Condensed Matter Phenomena*, Teaneck NJ, pp. 105–115. Singapore: World Scientific Publishing (1990).

Statistical theory of electron densities – multiple-scattering perturbation theory. *Proc. Roy. Soc. London A* **435**, 245–255 (1991).

Huang, D. M. and Chandler, D., Temperature and length scale dependence of hydrophobic effects and their possible implications for protein folding. *Proc. Natl. Acad. Sci. USA* **97**, 8324–8327 (2000).

Hummer, G., Garde, S., García, A. E., Paulaitis, M. E., and Pratt, L. R., Hydrophobic effects on a molecular scale. *J. Phys. Chem. B* **102**, 10 469–10 482 (1998*a*).

Hummer, G., Garde, S., Garcia, A. E., Pohorille, A., and Pratt, L. R., An information theory model of hydrophobic interactions. *Proc. Natl. Acad. Sci. USA* **93**, 8951–8955 (1996).

Hummer, G., Garde, S., García, A. E., and Pratt, L. R., New perspectives on hydrophobic effects. *Chem. Phys.* **258**, 349–370 (2000).

Hummer, G., Pratt, L., and García, A., Molecular theories and simulation of ions and polar molecules in water. *J. Phys. Chem. A* **102**, 7885–7895 (1998*b*).

Hummer, G., Pratt, L. R., and García, A. E., Hydration free-energy of water. *J. Phys. Chem.* **99**, 14 188–14 194 (1995).

Multistate gaussian model for electrostatic solvation free energies. *J. Am. Chem. Soc.* **119**, 8523–8527 (1997).

Imai, T. and Hirata, F., Partial molar volume and compressibility of a molecule with internal degrees of freedom. *J. Chem. Phys.* **119**, 5623–5631 (2003).

Jackson, J. D., *Classical Electrodynamics*. 2nd edition. New York: John Wiley & Sons (1975).

Jackson, J. L. and Klein, L. S., Potential distribution method in equilibrium statistical mechanics. *Phys. Fluids* **7**, 228–231 (1964).

Jaffe, R. L., Smith, G. D., and Yoon, D. Y., Conformation of 1,2-dimethoxyethane from *ab initio* electronic structure calculations *J. Phys. Chem.* **97**, 12745–12751 (1993).

Jaynes, E. T., The evolution of Carnot's Principle. In G. J. Erickson and C. R. Smith (eds.), *Maximum-Entropy and Bayesian Methods in Science and Engineering*, volume 1. Dordrecht: Kluwer (1988).

Probability Theory. The Logic of Science. Cambridge: Cambridge University Press (2003).

Kalos, M. H. and Whitlock, P. A., *Monte Carlo Methods, Volume I: Basics*. New York: Wiley-Interscience (1986).

Karplus, M. and Porter, R. N., *Atoms and Molecules: An Introduction for Students of Physical Chemistry*. New York: Benjamin (1970).

Kipnis, A. Y., Yavelov, B. E., and Rowlinson, J. S., *Van der Waals and Molecular Sciences*. New York: Oxford University Press (1996).

Kirkwood, J. G., Order and disorder in binary solid solutions. *J. Chem. Phys.* **6**, 70 (1938).

Statistical mechanics of cooperative phenomena. *J. Chem. Phys.* **8**, 623–627 (1940).

Kirkwood, J. G. and Poirier, J. C., The statistical mechanical basis of the Debye–Hückel theory of strong electrolytes. *J. Phys. Chem.* **86**, 591–596 (1954).

Kirkwood, J. G. and Salsburg, Z. W., The statistical mechanical theory of molecular distribution functions in liquids. *Disc. Faraday Soc.* **15**, 28–34 (1953).

Kirkwood, J. G. and Buff, F. P., The statistical mechanical theory of solutions. I. *J. Chem. Phys.* **19**, 774–777 (1951).

Klots, C. E., Solubility of protons in water. *J. Phys. Chem.* **85**, 3585–3588 (1981).

Kresse, G. and Furthmüller, J., Efficient iterative schemes for *ab initio* total-energy calculations using a plane-wave basis set. *Phys. Rev. B.* **54**, 11 169 (1996).

Kresse, G. and Hafner, J., *Ab initio* molecular dynamics for liquid metals. *Phys. Rev. B.* **47**, RC558 (1993).

Kubo, R., Generalized cumulant expansion method. *J. Phys. Soc. Jpn.* **17**, 1100 (1962).

Kumar, S. K., Szleifer, I., Hall, C. K., and Wichert, J. M., Computer simulation study of the approximations associated with the generalized Flory theories. *J. Chem. Phys.* **104**, 9100–9110 (1996).

Landau, L. D. and Lifshitz, E. M., *Statistical Physics*, from the preface to the first English edition. New York: Addison-Wesley (1969).

Course in Theoretical Physics: Electrodynamics of Continuous Media, volume 8. Oxford: Pergamon Press (1975).

Landau, L. D., Lifshitz, E. M., and Pitaevskii, L. P., *Course in Theoretical Physics: Statistical Physics, Part 1*, volume 5, 3rd edition. Oxford: Pergamon Press (1980).

Landu, L. D., Lifshitz, E. M., and Pitaevskii, L. P., *Electrodynamics of Continuous Media*, New York: Pergamon (1984), 2nd Edition.

LaViolette, R. A., Copeland, K. L., and Pratt, L. R., Cages of water coordinating Kr in aqueous solution. *J. Phys. Chem. A* **107**, 11 267–11 270 (2003).

Lebowitz, J. L. and Percus, J. K., Long-range correlations in a closed system with applications to nonuniform fluids. *Phys. Rev.* **122**, 1675 (1961*a*).

Thermodynamic properties of small systems. *Phys. Rev.* **124**, 1673 (1961*b*).

Asymptotic behavior of radial distribution function. *J. Math. Phys.* **4**, 248 (1963).

Lebowitz, J. L., Percus, J. K., and Verlet, L., Ensemble dependence of fluctuations with application to machine computations. *Phys. Rev.* **153**, 250 (1967).

Lebowitz, J. L., Stell, G., and Baer, S., Separation of interaction potential into two parts in treating many-body systems. I. General theory and applications to simple fluids with short-range and long-range forces. *J. Math. Phys.* **6**, 1282 (1965).

Lebowitz, J. L. and Waisman, E. M., Statistical-mechanics of simple fluids: beyond van der Waals. *Physics Today* **33**, 24–30 (1980).

Lee, B., Isoenthalpic and isoentropic temperatures and the thermodynamics of protein denaturation. *Proc. Natl. Acad. Sci. USA* **88**, 5154 (1991).

Leeuw, S. W. D., Perram, J. W., and Smith, E. R., Simulation of electrostatic systems in periodic boundary conditions. I. Lattice sums and dielectric constants. *Proc. Roy. Soc. Lond. A* **373**, 27–56 (1980).

Lefebvre, K. K., Lee, J. H., Balsara, N. P., *et al.*, Relationship between internal energy and volume change of mixing in pressurized polymer blends. *Macromol.* **32**, 5460–5462 (1999).

Lemberg, H. L. and Stillinger, F. H., Central-force model for liquid water. *J. Chem. Phys.* **62**, 1677–1690 (1975).

Lenhoff, A. M., A natural infection: chemical engineering and molecular biophysics. *AICHE J.* **49**, 806–812 (2003).

Lesieur, M., *Turbulence in Fluids*. Dordrecht: Kluwer (1997).

Levy, R. M., Belhadj, M., and Kitchen, D. B., Gaussian fluctuation formula for electrostatic free-energy changes in solution. *J. Chem. Phys.* **95**, 3627–3633 (1991).

Lewis, G. N., Randall, M., Pitzer, K. S., and Brewer, L., *Thermodynamics*, see Section 23 New York: McGraw-Hill (1961).

Librovich, N. B. and Maiorov, V. D., Vibrational spectra of the $H_3O_2^-$ and $D_3O_2^-$ ions in aqueous and in deuteriated aqueous solutions of potassium and sodium hydroxides. *Russ. J. Phys. Chem.* **56**, 380–383 (1982).

Librovich, N. B., Sakun, V. P., and Sokolov, N. D., H^+ and OH^- ions in aqueous solutions. Vibrational spectra of hydrates. *Chem. Phys.* **39**, 351–366 (1979).

Lobaugh, J. and Voth, G. A., The quantum dynamics of an excess proton in water. *J. Chem. Phys.* **104**, 2056–2069 (1996).

Ma, S.-K., *Statistical Mechanics*. New York: World Scientific (1985).

MacKerell Jr., A. D., Brooks, B., Brooks III, C. L., *et al.*, CHARMM: the energy function and its parameterization with an overview of the program. In *Encyclopedia of Computational Chemistry*, pp. 271–277. Chichester: John Wiley & Sons (1998).

MacKinnon, R., Potassium channels. *FEBS Letts.* **555**, 62–65 (2003).

Madras, N. and Sokal, A. D., Nonergodicity of local, length-conserving Monte Carlo algorithms for the self-avoiding walk. *J. Stat. Phys.* **47**, 573–96 (1987).

Maduke, M. and Miller, C., A decade of ClC chloride channels: structure, mechanism, and many unsettled questions. *Ann. Rev. Biophys. Biomol. Struct.* **29**, 411–438 (2000).

Maibaum, L., Dinner, A. R., and Chandler, D., Micelle formation and the hydrophobic effect. *J. Phys. Chem. B* **108**, 6778–6781 (2004).

Mansoori, G. A. and Matteoli, E., *Fluctuation Theory of Mixtures*. New York: Taylor & Francis (1990).

Marchi, M., Sprik, M., and Klein, M. L., Calculation of the free energy of electron solvation in liquid ammonia using a path integral quantum Monte Carlo simulation. *J. Phys. Chem.* **92**, 3625–3629 (1988).

Marcinkeiwicz, J., Sur une propriété de la loi de Gauß. *Math. Z* **44**, 612 (1939).

Marcus, Y., *Ion solvation*. London: Wiley (1985).

Markovitz, H. and Zapas, L. J., High polymer physics. In H. L. Anderson (ed.), *A Physicist's Desk Reference*, pp. 210–225. New York: American Institute of Physics (1989).

Mathews, J. and Walker, R. L., *Mathematical Methods of Physics*. New York: Benjamin (1964).

Mayer, J. E. and Montroll, E., Molecular distribution. *J. Chem. Phys.* **9**, 2–16 (1941).

McCrackin, F. L., Weighting methods for Monte Carlo calculation of polymer configurations. *J. Res. NBS, Section B (Mathematics and Mathematical Physics)* **76B**, 193–200 (1972).

McNaught, A. D. and Wilkinson, A., *IUPAC Compendium of Chemical Terminology*, 2nd edition. Cambridge: Royal Society of Chemistry (1997).

McQuarrie, D. A., *Statistical Mechanics*, see Chapter 14. New York: Harper & Row (1976).

Mooij, G. C. and Frenkel, D., Numerical test of the generalized Flory and generalized Flory dimer theories. *J. Chem. Phys.* **100**, 6088–6091 (1994).

Mootz, D. and Stäben, D., Clathrate hydrates of tetramethylammonium hydroxide – new phases and crystal-structures. *Z. Naturforsch. B - Chem. Sci.* **47**, 263–274 (1992).

Morita, T. and Hiroike, K., A new approach to the theory of classical fluids. III. *Prog. Theo. Phys.* **25**, 537–578 (1961).

Münster, A., *Statistical Thermodynamics*, volume 1. Berlin: Springer-Verlag (1969). *Statistical Thermodynamics*, volume 2. Berlin: Springer-Verlag (1974).

Murphy, K. P., Privalov, P. L., and Gill, S. J., Common features of protein unfolding and dissolution of hydrophobic compounds. *Science* **247**, 559–561 (1990).

Neyman, J. and Scott, E. L., The distribution of galaxies. *Scientific American* p. 187 (September, 1956).

Okamoto, H., Monte Carlo study of systems of oligomers in two-dimensional spaces. *J. Chem. Phys.* **64**, 2686–2691 (1993).

Oppenheim, I., Single electrode potentials. *J. Phys. Chem.* **68**, 2959–2961 (1964).

Paddison, S. J., Pratt, L. R., Zawodzinski, T., and Reagor, D. W., Molecular modeling of trifluoromethanesulfonic acid for solvation theory. *Fluid Phase Equilibria* **151**, 235–243 (1998).

Paddison, S. J., Pratt, L. R., and Zawodzinski, T. A., Variation of the dissociation constant of triflic acid with hydration. *J. Phys. Chem. A* **105**, 6266–6268 (2001).

Paulaitis, M. E., Ashbaugh, H. S., and Garde, S., The entropy of hydration of simple hydrophobic solutes. *Biophys. Chem.* **51**, 349–357 (1994).

Paulaitis, M. E. and Pratt, L. R., Hydration theory for molecular biophysics. *Adv. Prot. Chem.* **62**, 283–310 (2002).

Pauli, W., Über den Zusammenhangen des Abschlusses der Electronengruppen im Atom mit der Komplexstruktur der Spectren. *Z. Physik* **31**, 765–783 (1925).

Peierls, R., *Surprises in Theoretical Physics*. Princeton: Princeton University Press (1979).

Peierls, R. E., Zur Theorie des Diamagnetismus von Leitungselectronen. *Z. Physik* **80**, 763–791 (1933).

Percus, J. K., The pair distribution function in classical statistical mechanics. In *The Equilibrium Theory of Classical Fluids*, pp. II33–II170, New York: Benjamin (1964).

Pettitt, B. M., A perspective on "Volume and heat of hydration of ions"– Born M (1920) *Z. Physik* **1**, 45. *Theo. Chem. Acc.* **103**, 171–172 (2000).

Pierotti, R. A., Scaled particle theory of aqueous and non-aqueous solutions. *Chem. Rev.* **76**, 717–726 (1976).

Pitzer, K. S., Self-ionization of water at high-temperature and the thermodynamic properties of the ions. *J. Phys. Chem.* **86**, 4704–4708 (1982).

Thermodynamics of sodium-chloride solutions in steam. *J. Phys. Chem.* **87**, 1120–1125 (1983).

Plishke, M. and Bergerson, B., *Equilibrium Statistical Physics*. Singapore: World Scientific (1994).

Pohorille, A. and Pratt, L. R., Cavities in molecular liquids and the theory of hydrophobic solubilities. *J. Am. Chem. Soc.* **112**, 5066–5074 (1990).

Pohorille, A. and Wilson, M. A., Excess chemical potential of small solutes across water-membrane and water-hexane interfaces. *J. Chem. Phys.* **104**, 3760–3773 (1996).

Pratt, L. R., Contact potentials of solution interfaces: phase equilibrium and interfacial electric fields. *J. Phys. Chem.* **96**, 25– 33 (1992).

Hydrophobic effects. In *Encyclopedia of Computational Chemistry*, pp. 1286–1294. Chichester: John Wiley & Sons (1998).

Molecular theory of hydrophobic effects: "She is too mean to have her name repeated". *Annu. Rev. Phys. Chem.* **53**, 409–436 (2002).

Theory in action: highlights in the theoretical division at Los Alamos 1943–2003: modeling and molecular theory of liquids. Technical Report LA-14000-H, Los Alamos National Laboratory (2003).

Pratt, L. R. and Ashbaugh, H. S., Self-consistent molecular field theory for packing in classical liquids. *Phys. Rev. E* **68**, 021 505 (2003).

Pratt, L. R. and Chandler, D., Theory of the hydrophobic effect. *J. Chem. Phys.* **67**, 3863 (1977).

Pratt, L. R. and Haan, S. W., Effects of periodic boundary-conditions on equilibrium properties of computer-simulated fluids. 1. Theory. *J. Chem. Phys.* **74**, 1864–1872 (1981*a*).

Effects of periodic boundary-conditions on equilibrium properties of computer-simulated fluids. 2. Application to simple liquids. *J. Chem. Phys.* **74**, 1873–1876 (1981*b*).

Pratt, L. R., Hoffman, G. G., and Harris, R. A., Statistical theory of electron-densities. *J. Chem. Phys.* **88**, 1818–1823 (1988).

Ground-state densities from electron propagators – optimized Thomas–Fermi approximation for short wavelength modes. *J. Chem. Phys.* **92**, 6687–6696 (1990).

Pratt, L., Hummer, G., and Garde, S., Theories of hydrophobic effects and the description of free volume 529. in complex liquids. In C. Caccamo, J.–P. Hansen, and G. Stell (eds.), *New Approaches to Problems in Liquid State Theory*, volume 529. NATO Science Series, pp. 407–420. Netherlands: Kluwer (1999).

Pratt, L. R. and LaViolette, R. A., Quasi-chemical theories of associated liquids. *Mol. Phys.* **94**, 909–915 (1998).

Pratt, L. R., LaViolette, R. A., Gomez, M. A., and Gentile, M. E., Quasi-chemical theory for the statistical thermodynamics of the hard-sphere fluid. *J. Phys. Chem. B* **105**, 11 662–11 668 (2001).

Pratt, L. R. and Pohorille, A., Theory of hydrophobicity: Transient cavities in molecular liquids. *Proc. Natl. Acad. Sci. USA* **89**, 2995–2999 (1992).

Hydrophobic effects from cavity statistics. In M. U. Palma, M. B. Palma-Vittorelli, and F. Parak (eds.), *Proceedings of the EBSA 1992 International Workshop on Water–Biomolecule Interactions*, pp. 261–268. Bologna: Soc. Italiana de Fisica (1993).

Hydrophobic effects and modeling of biophysical aqueous solution interfaces. *Chem. Rev.* **102**, 2671–2691 (2002).

Pratt, L. R. and Rempe, S. B., Quasi-chemical theory and implicit solvent models for simulations. In L. R. Pratt and G. Hummer (eds.), *Simulation and Theory of Electrostatic Interactions in Solution. Computational Chemistry, Biophysics, and Aqueous Solutions*, volume 492 of *AIP Conference Proceedings*, pp. 172–201. Melville, NY: American Institute of Physics (1999).

Predescu, C., The partial averaging method. *J. Math. Phys.* **44**, 1226–1239 (2003).

Press, W. H., Teukolsky, S. A., Vetterling, W. T., and Flannery, B. P., *Numerical Recipes in FORTRAN*, 2nd edition. Cambridge: Cambridge University Press (1992).

Privalov, P. L., Cold denaturation of proteins. *Crit. Rev. Biochem. Mol. Biol.* **25**, 281–305 (1990).

Privalov, P. L. and Gill, S. J., The hydrophobic effect: a reappraisal. *Pure Appl. Chem.* **61**, 1097 (1989).

Reiss, H., Superposition approximations from a variation principle. *J. Stat. Phys.* pp. 39–47 (1972).

Scaled particle theory of hard sphere fluids to 1976. In U. Landman (ed.), *Statistical Mechanics and Statistical Methods in Theory and Application*, pp. 99–138. New York: Plenum (1977).

Statistical geometry in the study of fluids and porous media. *J. Phys. Chem.* **96**, 4736–4747 (1992).

Reiss, H., Frisch, H. L., and Lebowitz, J. L., Statistical mechanics of rigid spheres. *J. Chem. Phys.* **31**, 369 (1959).

Rempe, S. B., Asthagiri, D., and Pratt, L. R., Inner shell definition and absolute hydration free energy of K^+(aq) on the basis of quasi-chemical theory and *ab initio* molecular dynamics. *Phys. Chem. Chem. Phys.* **6**, 1966–1969 (2004).

Rempe, S. B. and Pratt, L. R., The hydration number of Na^+ in liquid water. *Fluid Phase Equilibrium* **183–184**, 121–132 (2001).

Rempe, S. B., Pratt, L. R., Hummer, G., *et al.* The hydration number of Li^+ in liquid water. *J. Am. Chem. Soc.* **122**, 966–967 (2000).

Resnick, S. I., *A Probability Path*. New York: Birkhäuser (2001).

Riordan, J., *An Introduction to Combinatorial Analysis*. Princeton, NJ: Princeton University Press (1978).

Robertson, A. D. and Murphy, K. P., Protein structure and the energetics of protein stability. *Chem. Rev.* **97**, 1251–1267 (1997).

Robertson, W. H., Diken, E. G., Price, E. A., Shin, J. W., and Johnson, M. A., Spectroscopic determination of the OH^- solvation shell in the $OH^- \cdot (H_2O)_n$ clusters. *Science* **299**, 1367–1372 (2003).

Rosenbluth, M. N. and Rosenbluth, A. W., Monte Carlo calculation of the average extension of molecular chains. *J. Chem. Phys.* **23**, 356–359 (1955).

Roux, B. and Simonson, T., Implicit solvent models. *Biophys. Chem.* **78**, 1–20 (1999).

Rowlinson, J. S. and Swinton, F. L., *Liquids and Liquid Mixtures*. New York: Butterworths (1982).

Rowlinson, J. S. and Widom, B., *Molecular Theory of Capillarity*. Oxford: Clarendon (1982).

Rustad, J. R., Felmy, A. R., Rosso, K. M., and Bylaska, E. J., *Ab initio* investigation of the structures of NaOH hydrates and their Na^+ and OH^- coordination polyhedra. *Amer. Min.* **88**, 436–449 (2003).

Sauer, N. N., McCleskey, T. M., Taylor, T. P., *et al.*, Ligand associated dissolution of beryllium: toward an understanding of chronic beryllium disease. Technical report LA-UR-02-1986, Los Alamos National Laboratory (2002).

Schiöberg, D. and Zundel, G., Very polarisable hydrogen bonds in solutions of bases having infra-red continua. *J. Chem. Soc. Faraday Trans. II.* **69**, 771–781 (1973).

Schlijper, A. G. and Kikuchi, R., A variational approach to distribution function theory. *J. Stat. Phys.* **61**, 143–160 (1990).

Schmitt, U. W. and Voth, G. A., The computer simulation of proton transport in water. *J. Chem. Phys.* **111**, 9361–9381 (1999).

Schwegler, E., Grossman, J. C., Gygi, F., and Galli, G., Towards an assessment of the accuracy of density functional theory for first principles simulations of water. II. *J. Chem. Phys.* **121**, 5400–5409 (2004).

Shaikh, S. A., Ahmed, S. R., and Jayaram, B., A molecular thermodynamic view of DNA-drug interactions: a case study of 25 minor-groove binders. *Arch. Biochem. Biophys.* **429**, 81–99 (2004).

Shing, K. S. and Chung, S. T., Computer-simulation methods for the calculation of solubility in supercritical extraction systems. *J. Phys. Chem.* **91**, 1674–1681 (1987).

Shore, J. E. and Johnson, R. W., Axiomatic derivation of the principle of maximum-entropy and the principle of minimum cross-entropy. *IEEE Trans. Inf. Theory* **26**, 26–37 (1980).

Siepmann, J. I., Monte Carlo methods for simulating phase equilibria of complex fluids. *Adv. Chem. Phys.* **105**, 443–460 (1999).

Singer, A., Maximum entropy formulation of the Kirkwood superposition approximation. *J. Chem. Phys.* **121**, 3657–3666 (2004).

Singer, S. J. and Chandler, D., Free energy functions in the extended RISM approximation. *Mol. Phys.* **55**, 621–5 (1985).

Slusher, J. T. and Mountain, R. D., A molecular dynamics study of a reversed-phase liquid chromatography model. *J. Phys. Chem. B* **103**, 1354–1362 (1999).

Smith, P. E., Computer simulation of cosolvent effects on hydrophobic hydration. *J. Phys. Chem. B* **103**, 525–534 (1999).

Sokal, A. D., Monte Carlo methods for the self-avoiding walk. In K. Binder (ed.), *The Liquid State of Matter: Fluids, Simple and Complex*, pp. 47–124, New York: Oxford University Press (1995).

Stakgold, I., *Green's Functions and Boundary Value Problems*, see, for example, Section 4.1, New York: Wiley-Interscience (1979).

Stell, G., Cluster expansions for classical systems in equilibrium. In *The Equilibrium Theory of Classical Fluids*, pp. II171–II266, New York: Benjamin (1964).

  Fluids with long-range forces. In B. J. Berne (ed.), *Statistical Mechanics. Part A: Equilibrium Techniques*, pp. 47–84, New York: Plenum (1977).

  Mayer–Montroll equations (and some variants) through history for fun and profit. In M. F. Shlesinger and G. H. Weiss (eds.), *The Wonderful World of Stochastics: A Tribute to Elliot W. Montroll*, volume XII of *Studies in Statistical Mechanics*, pp. 127–156. New York: Elsevier Science Publishers (1985).

Stillinger, F. H., Structure in aqueous solutions of nonpolar solutes from the standpoint of scaled-particle theory. *J. Soln. Chem.* **2**, 141–158 (1973).

  Proton transfer reactions and kinetics in water. In H. Eyring and D. Henderson (eds.), *Theoretical Chemistry: Advances and Perspectives*, volume 3, pp. 177–234. New York: Academic (1978).

  Low frequency dielectric properties of liquid and solid water. In *The Liquid State of Matter: Fluids, Simple and Complex*, pp. 341–431, Amsterdam: North-Holland (1982).

Tawa, G. J. and Pratt, L. R., Tests of dielectric model descriptions of chemical charge displacements in water. *ACS Symp. Series* **568**, 60–70 (1994).

  Theoretical calculation of the water ion product K_W. *J. Am. Chem. Soc.* **117**, 1625–1628 (1995).

Thiele, E., Equation of state for hard spheres. *J. Chem. Phys.* **39**, 474 (1963).

Thompson, J. and Begenisich, T., External TEA block of shaker K^+ channels is coupled to the movement of K^+ ions within the selectivity filter. *J. Gen. Physiol.* **122**, 239–246 (2003).

Thuraisingham, R. A. and Friedman, H. L., HNC solution for the central force model for liquid water. *J. Chem. Phys.* **78**, 5772–5775 (1983).

Tildesley, D. J. and Streett, W. B., An equation of state for hard-dumb-bell fluids. *Mol. Phys.* **41**, 85–94 (1980).

Tomberli, B., Benmore, C. J., Egelstaff, P. A., Neuefeind, J., and Honkimaki, V., Isotopic quantum effects in water structure measured with high energy photon diffraction. *J. Phys. – Cond. Matt.* **12**, 2597–2612 (2000).

Tomberli, B., Egelstaff, P. A., Benmore, C. J., and Neuefeind, J., Isotopic quantum effects in the structure of liquid methanol: I. experiments with high-energy photon diffraction. *J. Phys. – Cond. Matt.* **13**, 11 405–11 420 (2001).

Torrie, G. M. and Valleau, J. P., Nonphysical sampling distributions in Monte Carlo free-energy estimation: umbrella sampling. *J. Comp. Phys.* **23**, 187–199 (1977).

Triolo, R. and Narten, A. H., Diffraction pattern and structure of aqueous hydrochloric acid solutions at 20 °C. *J. Chem. Phys.* **63**, 3624–3631 (1975).

Tuckerman, M. E., Marx, D., and Parrinello, M., The nature and transport mechanism of hydrated hydroxide ions in aqueous solution. *Nature* **417**, 925– 929 (2002).

Uematsu, M. and Franck, E. U., Static dielectric-constant of water and steam. *J. Phys. Chem. Rev. Data* **9**, 1291–1305 (1980).

Uhlenbeck, G. E. and Ford, G. W., *Lectures in Statistical Mechanics*. Providence: American Mathematical Society (1963).

Ursell, H. D., The evaluation of Gibbs' phase integral for imperfect gases. *Proc. Cambridge Philos. Soc.* **23**, 685 (1927).

Valleau, J. P. and Cohen, L. K., Primitive model electrolytes. I. Grand canonical Monte Carlo computations. *J. Chem. Phys.* **72**, 5935–5941 (1980).

Valleau, J. P. and Torrie, G. M., A guide to Monte Carlo for statistical mechanics: 2. Byways. In *Statistical Mechanics, Part A: Equilibrium Techniques*, pp. 169–194. New York: Plenum (1977).

Van Kampen, N. G., *Stochastic Processes in Physics and Chemistry*, see Section II.1. Amsterdam: North-Holland (1992).

Vuilleumier, P. and Borgis, D., Transport and spectroscopy of the hydrated proton: a molecular dynamics study. *J. Chem. Phys.* **111**, 4251–4266 (1999).

Wang, Q., Johnson, J. K., and Broughton, J. Q., Path integral grand canonical Monte Carlo. *J. Chem. Phys.* **107**, 5108–5117 (1997).

Weeks, J. D., Connecting local structure to interface formation: a molecular scale van der Waals theory of nonuniform liquids. *Annu. Rev. Phys. Chem.* **53**, 533–562 (2002).

Weingartner, H., Klante, D., and Schneider, G. M., High-pressure liquid–liquid immiscibility in aqueous solutions of tetra-n-butylammonium bromide studied by a diamond anvil cell technique. *J. Soln. Chem.* **28**, 435–446 (1999).

Weingartner, H. and Steinle, E., *p, T, x* surface of liquid–liquid immiscibility in aqueous solutions of tetraalkylammonium salts. *J. Phys. Chem.* **96**, 2407–2409 (1992).

Wertheim, M. S., Exact solution of Percus–Yevick integral equation for hard spheres. *Phys. Rev. Lett.* **10**, 321 (1963).

Widom, B., Some topics in the theory of fluids. *J. Chem. Phys.* **39**, 2808–2812 (1963).

Random sequential addition of hard spheres to a volume. *J. Chem. Phys.* **44**, 3888–3894 (1966).

Intermolecular forces and the nature of the liquid state. *Science* **157**, 375–382 (1967).

Structure of interfaces from uniformity of the chemical potential. *J. Stat. Phys.* **19**, 563–574 (1978).

Potential-distribution theory and the statistical mechanics of fluids. *J. Phys. Chem.* **86**, 869–872 (1982).

Statistical Mechanics. A Concise Introduction for Chemists. Cambridge: Cambridge University Press (2002).

Wood, W. W., Early history of computer simulations in statistical mechanics. In *International School of Physics "Enrico Fermi,"* volume XCVII, pp. 3–14. Bologna: Soc. Italiana di Fisica (1986).

Some additional recollections, and the absence thereof, about the early history of computer simulations in statistical mechanics. In *Euroconference on Computer Simulation in Condensed Matter Physics and Chemistry*, volume 49, pp. 908–911. Bologna: Soc. Italiana di Fisica (1996).

Xu, H., Freidman, H. L., and Raineri, F. O., Electrolyte solutions that unmix – hydrophobic ions in water. *J. Soln. Chem.* **20**, 739–773 (1991).

Yethiraj, A. and Hall, C. K., Generalized Flory equations of state for square-well chains. *J. Chem. Phys.* **95**, 8494–8506 (1991).

Yin, J., Kuang, Z., Mahankali, U., and Beck, T. L., Ion transit pathways and gating in ClC chloride channels. *Proteins: Struct., Funct., and Bioinform.* **57**, 414–421 (2004).

Yoon, B. J. and Lenhoff, A. M., A boundary element method for molecular electrostatics with electrolyte effects. *J. Comp. Chem.* **11**, 1080–1086 (1990).

Zatsepina, G. N., State of the hydroxide in water and aqueous solutions. *Zhur. Struk. Khim. (English edn.)* **12**, 894–898 (1971).

Zhang, L., Sun, L., Siepmann, J. I., and Schure, M. R., Molecular simulation study of the bonded-phase structure in reversed-phase liquid chromatography with neat aqueous solvent. *J. Chromatogr. A* **1079**, 127–135 (2005).

Zhou, R. H., Huang, X. H., Margulis, C. J., and Berne, B. J., Hydrophobic collapse in multidomain protein folding. *Science* **305**, 1605–1609 (2004).

Zhou, Y. Q., Stell, G., and Friedman, H. L., Note on standard free energy of transfer and partitioning of ionic species between two fluid phases. *J. Chem. Phys.* **89**, 3836–3839 (1988).

Zhu, Z. and Tuckerman, M. E., *Ab initio* molecular dynamics investigation of the concentration dependence of charged defect transport in basic solutions via calculation of infrared spectrum. *J. Phys. Chem. B* **106**, 8009–8018 (2002).

Ziman, J. M., *Principles of the Theory of Solids*. New York: Cambridge University Press (1972).

Zuckerman, D. and Woolf, T., Overcoming finite-sampling errors in fast-switching free-energy estimates: extrapolative analysis of a molecular system. *Chem. Phys. Letts.* **351**, 445–453 (2002).

Zwanzig, R. W., High-temperature equation of state by a perturbation method. I. Nonpolar gases. *J. Chem. Phys.* **22**, 1420–1426 (1954).

Index